A Desert Calling

Michael A. Mares

A Desert Calling

LIFE IN A
FORBIDDING
LANDSCAPE

HARVARD UNIVERSITY PRESS
Cambridge, Massachusetts, and London, England
2002

Photo on p. ii by M. A. Mares: a vicuña on a sand dune against a backdrop
of an arid Andean slope, Catamarca Province, Argentina.

Library of Congress Cataloging-in-Publication Data
Mares, Michael A.
A desert calling : life in a forbidding landscape / Michael A. Mares.
p. cm.
Includes bibliographical references (p.).
ISBN 0-674-00747-6 (hardcover : alk. paper)
1. Mammals.   2. Desert animals.   3. Biology—Field work.   I. Title.
QL703 .M368 2002
599.1754—dc21   2001051786

For Lynn, Gabriel, and Danny, the fixed stars in my desert sky

Contents

Foreword

BY STEPHEN JAY GOULD

I have long been amused that the "standard" desert island cartoon shows a man, perhaps leaning against a lone palm tree, but inevitably situated on a small featureless space fully covered in sand—with the clear (and mistaken) inference that the "desert" in desert island has the same root and meaning as the Sahara. Same root yes, but entirely different meaning, hence my amusement at the error. As Michael Mares points out, the Latin *desertus* means solitary or forsaken—so desert islands may be lush with vegetation, and humid as could be. Only the lone and shipwrecked seafarer is forsaken and deserted there. But the Sahara and other dry and sandy places are misnamed for a different fallacy based on the same root. We think that they are lifeless and forsaken because we can't take the heat, the absence of the vegetation we know best, and especially the dangerous scarcity of water. Yet, as they say in Jurassic Park, life finds a way.

You may have to look a little harder (and Mares tells some funny stories about his inability to trap mammals in the South American desert until he figured out where they were and what they liked). But deserts teem with life amidst, around, and beneath the surface of these visually bleakest places on earth. And since rodents are, by far, the most abundant and diverse of mammalian groups, we should not be surprised that if any representatives of our "highest" class have managed to prevail in such places, rodents should be the prime candidates. Indeed, by managing to thrill and delight us with a conjunction of the two absolutely least promising subjects for human fascination—rats for creatures and deserts for places—Mares has truly proven, in this dual symbolic utility, that we live on a most wondrous planet, a place where absolutely every nook of space and every sentient object (even the insentient ones, for that matter) loudly proclaim the truth of Shakespeare's appended examples for his famous proclamation about the sweet uses of adversity: "sermons in stones, and good in every thing."

Michael Mares, the world's expert on the natural history of desert rodents, writes from a perspective and life experience that will immediately resonate with the worldview and practice of any professional natural historian, but that will seem unusual, even exotic or a bit crazy, to many others. Two passions regulate the activities of this wonderful world. First, a love and spirited defense for the most maligned of professional activities in science: taxonomy, or the classification of organisms, so often, and ignorantly, seen as a farrago of Latin mumbo-jumbo applied by beancounters and bookkeepers interested only in placing objects into their appropriate pigeon holes (or rodent holes in this case). But, in nature's ecology, sexual organisms group themselves into species, and species thereby become the true and basic units of biodiversity, the framework of life on our planet. Yes, species require names, just as individual humans want and need distinctive labels. But naming per se (and in Latin) is only a tool for pursuing the real work of understanding how populations of organisms exist, adapt, thrive, die, and interact in nature. Through this book we come to understand why Mares and his colleagues must identify and name the units of life in order to understand the ecology of the desert—and we learn how far astray we can go (often with practically disastrous consequences, particularly in medicine and agriculture) when we haughtily ignore taxonomy, disregard the small but distinctive differences among real species, fail to study living organisms in their natural habitats, and falsely assume that all rodents looking basically alike, and coming from the same broad region, must be the "same" animal. And we learn to appreciate why Mares designates Oldfield Thomas as his hero, the man who named nearly three thousand species and subspecies of mammals, and wrote such a touching afterword, in genuine humility before nature's vast diversity, to the paper that named his two-thousandth species.

Second, and even more distinctively, Mares's first love, and persistently dominant passion, lies in fieldwork, often under the most appalling and quite dangerous conditions, especially in deserts, with utterly unsuitable vehicles and not a single gas station between Marrakesh and Timbuktu. Fieldworkers are an odd breed—and I say this as someone who has more than merely dipped into this world, but still lives basically outside it, thereby gaining both enough experience to understand the allure and enough distance to recognize the peculiarity. To illustrate the power of Mares's love and true obsession, I need only mention that he devotes virtually every word of this autobiographical book to loving (and sometimes gruesome) details of all his field trips, but then grants only part of a single paragraph to mentioning his genuine "day job" of the last several years—his brilliantly successful directorship of the new and stunning Sam Noble Oklahoma Museum of Natural History at the University of Oklahoma. Now here's a man who knows what's important in natural history!

Field narratives have certain conventions, and Mares follows them here, but with a verbal freshness (and a fine sense for a good yarn) that will delight even the most sophisticated urbanite who views Montauk Point as the ultimate wilderness. Of the two major desiderata of the genre, one must first tell terrific stories about animals—as Mares does again and again. Each reader will have a personal favorite. I was stunned by how the plains vizcacha rat of Argentina eats saltbush, when the salt crystals must first be scraped away to reach the edible leaf underneath. Two stiffened bundles of hair, shaped just like teeth, "articulate" with the lower incisors (true teeth) to chisel away the salt, which flies from the animal's mouth in all directions. Mares comments: "Few mammals use hair to assist in gathering food. The most ready examples are the baleen whales, those massive oceanic beasts that have developed filters of modified hair—the baleen—that strain microorganisms from the sea as the whale pumps ocean water through the mouth. In one respect, the little plains vizcacha rat is a whale of a rodent."

Second, one must relate the tales of danger, biting bugs, venomous snakes, near drowning, strandings in the desert, and meetings with weird and dangerous people—the occasional but inevitable incidents that no one really loves when they are happening, but that more than repay the debt in the pleasure of later telling. In this category, my personal favorite also leaves me duty bound to chastise my old buddy Mike Mares. He meets a grizzled codger in the most godforsaken and isolated spot of the Argentine desert. The guy claims he's from Detroit so Mares springs the trap and talks to him in English. But the guy comes up golden, speaking our native lingo perfectly. Then he tells Mares his story: He's been living in the desert some forty years and lost his passport about thirty years ago. His brother, he says, had been in Al Capone's gang, but left to set up a rival organization. Capone threatened to kill him, so the whole family had to run as far away as possible. And Mares swallows it.

Now c'mon Mike; I'll buy the Detroit thing, but my one will get you ten if anyone in his family ever spoke to Al Capone. People end up in the weirdest places for the oddest reasons (and the absolute best of fabricated stories). You always run into someone like Sam from Detroit, anywhere in the world. To tell my own politically incorrect tale, I was once working on an isolated island, population about a hundred native Bahamians, not a facility anywhere, and nary a visitor for months or years. So I'm collecting snails along the beach with my research assistant and I hear someone call out in an obvious American accent: "White people! What the hell are you doing here!" He had some story about his boat and his adventures all over the world, but if he could ever have reached the next island in that tub, I'd be surprised. Only white man there for years, he said. Maybe he first went there to escape Tony Soprano. Maybe he was the same guy that Mares met in Argentina.

Finally, because each stone does preach a sermon, books on natural history can only, as this book does so well, transcend the genre of a transient set of good yarns by tying these particulars to deep, important, and general problems of science that form the legitimate excuse for people putting themselves in danger, and spending so much time and money (though the latter sure seems scarcer than the former to field biologists), to pursue these private passions. In Mares's case, evolutionary theory sets this proper and general context, and Mike has masterfully woven his narrative of disparate stories around the central theoretical question that has motivated all his work—in one sense the best and deepest question of all the sciences of natural history: how much of life's diversity falls into predictable patterns regulated by scientific laws of evolution and biomechanics, and how much records the particular happenstances of singular places (in other words, the sensible, but unrepeatable, working out of unique historical sequences of events).

Evolutionists test this great question in the way that any experimental scientist would, although we must search for "experiments" performed for us by nature: by looking for independent replications under circumstances as identical as possible. In evolutionary terms, we ask: if faunas evolve independently in different parts of the world, but in climates and ecologies of maximal similarity, will we find the same adaptations, the same ecological strategies, repeated again and again by the historically different inhabitants of each region? If only a few ways of making a living can work in such a harsh place as a desert, will these modes be evolved again and again in different places—a phenomenon called "convergence," as illustrated, for example, in the separately evolved, but aerodynamically so similar, wings of bats, birds, and the extinct pterosaurs, or flying reptiles of dinosaur times. After all, flying isn't easy, and can only work in a few basic ways when you have to use bone and muscle, rather than oil and steel.

Now Mares and I have different suspicions about the relative weights of convergence versus unique historical oddity. I find the latter more fascinating and portentous; he thrills to the former. But we share the sense of all evolutionists that both modes make powerful contributions to the wondrously sensible diversity of life, and that no question could be more important to our field. Now Mike is hung up on one particular question amidst this generality—the best application to his field of desert rodents. I therefore end this foreword simply by honoring the importance of such private passions—and the beauty of genuine, factual, and fascinating resolutions sometimes provided by recalcitrant nature—in the best way I know: by retelling the lovely story of Mike's very best piece of detective work.

Michael Mares has put himself into great discomfort and occasional danger in nearly every great desert of the world, largely because he is consumed by a personal quest to understand and document one of the great convergences in biology: the propensity of some rodents (with the American kangaroo rat as the best-known local example) to become bipedal (two-legged) in deserts, where good biomechanical arguments identify hopping as an excellent mode of locomotion for many habitats in such places. Mike had to face a problem of great personal salience for a biologist of Hispanic background in the Americas: why, alone among the world's appropriate habitats, do the great South American deserts of Argentina and Chile lack any bipedal species? At first, he doesn't believe that the claim can be true, for these deserts have been so underexplored for rodents. So he devotes years to the search for a bipedal species, and finds nothing.

Then he develops a perfectly good hypothesis, but based upon his unfavored alternative of real difference for unique historical reasons. The native rodents of South America all belong to an odd group, called the caviomorphs, and including the monsters of the rodent world, from the coypu to the pig-sized capybara. Perhaps, by some quirk of a particular evolutionary past, the caviomorphs simply lack the wherewithal to evolve a two-legged lifestyle, even though the adaptation would be advantageous for some desert forms. Moreover, the "ordinary" rodents—that is, the species belonging to the conventional group of mice, rats, squirrels, and their ilk on all other major continents—may not have evolved a bipedal form simply because they haven't been in the southern desert long enough. South America was an island continent until the Isthmus of Panama rose just two to three million years ago—and no "ordinary" rodent got to southern South America as a northern migrant before then.

But just as he is beginning to feel comfortable with this unloved hypothesis, Mares snatches victory from the jaws of defeat in the quirkily and utterly unexpected way of virtually all major discovery. (After all, one cannot go out to look actively for the unexpected). He is casually reading, one day in 1973, a paper on South American paleontology by the greatest professional in the field: George Gaylord Simpson. And he finds his bipedal forms—but with a wonderful twist that validates both the odd contingency of peculiar circumstances and the faith that convergence grants some predictability to evolution. First of all, these bipedal desert forms are extinct. Second, they weren't rodents at all, but a group of marsupials that lived in the habitat of rodents and served as their ecological "vicar" in isolated South America. (Just as on the island continent of Australia, several marsupial groups evolved and flourished in the separateness of South America. Some have survived, including the misnamed "Virginia" opossum—a South American migrant after the rise of the Isthmus. But many others died after the mixture of faunas that followed the joining of South America to the rest of the world. Most notably, the large native mammalian carnivores of South America were all marsupials, and they are all gone. The jaguar evolved from another North American migrant.)

I thus end with Mike Mares's lovely resolution to the motivating scientific problem of his life as a field naturalist and student of living animals—solved by reading a paper about fossils: "The animals had disappeared in the Pleistocene, only about a million years ago, and had inhabited the Monte Desert near Andalgalá. Here was my bipedal desert specialist of the Monte Desert, only instead of being a rodent it was a marsupial . . . If I had arrived in Argentina a million years ago, instead of in 1970, I would have caught them in my traps and seen at once that they were strong ecological equivalents of the classical desert rodents of the world. I had simply gotten there too late."

When I go back, as I must, to live in a world almost wholly man-made and almost wholly absorbed in problems which man himself has created, I shall often return in memory to things seen and done during my desert interlude. There will be, first of all, encounters with birds and beasts to be remembered. Then, as on a screen, I shall see my mental kodachromes projected—sometimes of vast vistas of mountain or plain, sometimes little close-ups of an improbable blossom bursting out of a cactus, or of a lizard poised for a moment in the sun. All too often, I am afraid, I shall be reminded how whole acres of New York City in which nothing grows have been turned into a desert far more absolute than any I have ever seen in the Southwest, and I shall wonder whether man himself can live well in a place where nothing else can live at all. But I doubt whether anything else will be so continuously in the back of my mind as the consciousness of that metaphor which two thousand miles of countryside set forth, and I shall not forget its lesson: much can be lacking in the midst of plenty; on the other hand, where some things are scarce others, no less desirable, may abound.

Joseph Wood Krutch, *Desert Year*, 1985

The tropical Namib Desert, Namibia. (Photo by M. A. Mares.)

Prologue

Many of us grow up now with romantic notions about cactus-studded landscapes lit by neon sunsets, notions some scholars think may be the emerging landscape stereotype of the American West in the eyes of the world. But of course you know in your heart that the desert aesthetic must be a very recently acquired taste, for its opposite—a revulsion and even fear at its barrenness and screaming heat—flirts about the edges of the mind.

Dan Flores, *Horizontal Yellow*, 1999

I have devoted much of my career to studying mammals in deserts, investigating how they adapt to aridity, how they come together to form complex communities of organisms, and how they manage to exist in habitats of such singular starkness. Most people, however, associate the word "desert," whose Latin root is *deserare*, meaning abandoned, with a place that is barren—a wasteland, an area of magnificent desolation, harsh to all forms of life, especially human beings. And indeed every year people perish in the desert when caught unprepared for life in a hot, arid place. They have car trouble or get lost, having entered the desert with little or no water. They were unaware that in the desert the human organism requires 2 gallons of water each day to remain in water and temperature balance.

With too little water people perspire to maintain body temperature, but as water is lost blood thickens and salt levels in the tissues diminish. When salt and water levels are reduced, the body stops sweating, thereby reducing water loss, but increasing body temperature. If salt and water are not consumed, heatstroke, perhaps hallucinations, and death soon occur. The careless traveler who enters the desert with insufficient water and little or no salt becomes one with the desert, a choice bit of food and water for those many organisms—from the tiniest desert shrew to the African lion—that, through their remarkable array of behavioral, physiological, and ecological adaptations, do not find the desert particularly challenging.

If you are a field biologist working alone at a remote desert site there is no one to tell when or where you will be at some particular point, for you may be gone for weeks or months. As Clint Eastwood said, one has to know one's limitations. Cars break down. Accidents happen. Falls occur; legs are broken. Venomous snakes are a small but ever present danger, and in some areas even killer bees may pose a risk.

Many things can go wrong on a field trip across the desert, and the more remote the study area the greater the danger. Some bad things that can happen cannot be foreseen. But one thing a desert biologist knows: water and salt are the most important things to take into the desert. You can survive without food for up to a month, but without water you can live for only a very few days. The length of a field trip is often determined by water limitations, not by the amount of food or fuel remaining. Push the limits on water in a desert and you push the limits on life.

Clearly, deserts pose great challenges to people who enter them. Yet deserts have always been a part of humanity. Humans are not a species of the verdant jungle. We developed our most human characteristics in a dry savanna only marginally more luxuriant than a desert. It is the need to organize families, tribes, and societies in the face of an unpredictable environment that helped form societal structures over the millennia. The Egyptians of Pharaonic times were a desert people subsisting along the green thread of the Nile within the vast Sahara Desert. Though the Nile supplied water for them and their crops and animals, it was the desert that made them reach beyond the verdant valley toward other lands, and it was the desert that sup-

Broad vista across the Negev Desert, Israel, where Moses and the Israelites wandered for forty years. (Photo by M. A. Mares.)

plied the minerals and other materials the Egyptians needed to create a great civilization.

Deserts, with their clear air and limitless vistas, taught people to organize, for desert survival requires organization. Indeed, the harsh surrounding desert that placed a limit on the green oases of life may have permitted the growth of dense human populations within the circumscribed area of the Fertile Crescent. Increased population density may well have been necessary for the rise of civilization. Humanity's first system of codified laws also was created in the Cradle of Civilization—the valley of the Tigris and Euphrates rivers—where agriculture first was practiced in a semiarid habitat. Water, the sine qua non of desert life, was in short supply, and people struggled to devise fair ways of using this life-giving liquid. Crops, lives—civilization itself—depended on agriculture, which depended on water.

The spirit of the desert is set deep within the fiber of our being as a species. Yet although deserts made us what we are today, we have escaped those early xeric connections. We worry less about the unpredictable ways of nature than we did when we were peoples of the parched land. We no longer understand the stark, spare landscape of the desert.

But despite the popular view of the desert as a lifeless wilderness, and despite our own fragile ability to exist in an arid climate, life in the desert is, in fact, rich and mysterious, varied and complex. Landscapes that seem to harbor few organisms can suddenly burst forth with flowers within days of a rain shower. Dry waterholes can fill with shrimp and frogs overnight following a thunderstorm. A panorama of barren, burning dunes during the day can when darkness falls become a dramatic stage for predator and prey, as beetles, lizards, rodents, snakes, armadillos, marsupials, and even lions and elephants move across sand cooled by the desert night. As the sun rises, the life-filled desert again assumes the guise of a forbidding landscape, its many species awaiting the passing of the day.

The Search for Undiscovered Life

Most biological diversity, . . . in the old-fashioned way, awaits discovery by foot, net, and scuba gear. To confront diversity, biologists continue to go out of the laboratory and into the world . . . they press on into unexplored mountain ridges, river headwaters, and coral reefs. For most countries in the world . . . the plumb line is still being let out; we have no idea where it will all end. The rewards of physical adventure, the excitement of grimy, sweat-soaked exploration into remote corners of the earth, still beckon in science.

Edward O. Wilson, *The Diversity of Life*, 1992

In this book I describe what it is like to do field research on mammals throughout the world. I am one of the army of people who make the millions of observations on animals that form the backbone of the scientific data of nature. Field biologists like me are the foot soldiers of natural history, scientists whose passion for going into the field to study organisms provides the foundational data for understanding nature and, indeed, life itself.

The animals that I study are not as charismatic as the African lions described so effectively by Craig Packer in *Into Africa*. I have focused instead on the uncharismatic, the rodents and other small mammals that pass remarkable lives mostly hidden from view, perhaps unobserved in a desert or unstudied in a rainforest. Studying these species in the field for more than thirty years has provided me with many insights into the astonishing variety and diversity of life. My time in the field has helped me to understand how species of small mammals adapt to, and manage to persist in, some of the most hostile habitats on earth, and has given me a singular appreciation for the tenacity and ubiquity of life.

Field biology involves daily encounters with the wondrous. There are no limits to nature's marvelous diversity of life and the subtle interactions that bind species together. Whether predator and prey or parasite and host, all creatures are immutably concatenated in the slow dance of evolution across time. There are downsides to field research, to be sure. Being a field biologist makes having a home life difficult. I long for home while in the field, but yearn to be in the field when I'm at home. Field biologists, especially those who work in foreign countries, must love their discipline very much for they immerse themselves in fieldwork for months or even years at a time, willingly risking disease, accidents, bureaucratic snafus, and other difficulties. It is a discipline that demands a special dedication.

My field crew in the high Andes of Mendoza Province, Argentina. An undescribed genus and species of mammal was found at this locality. (Photo by M. A. Mares.)

To do field research is to lead an uncommon life that is continuously renewed by nature. I have traveled to all the states in the United States, dozens of countries, and all continents except Antarctica. I have studied mammals in rainforests, deserts, mountains, and grasslands. To be in the field, I have lived and worked in dictatorships, monarchies, and democracies as I pursued my research on small mammals—rodents, bats, and other cryptic inhabitants of the less frequented parts of the world. My work has required that I leave my family, home, and country for long periods of time, and in this my experiences are typical of other biologists who have gone into the field in search of new insights into little known species.

My small field crew in Patagonia, Chubut Province, Argentina. The Land Rover served as a break for the incessant wind, which eventually destroyed all the tents. (Photo by M. A. Mares.)

Different field biologists conduct research in different ways. Some travel with enormous retinues: many people, much equipment, and many vehicles. They are like an army, attacking a research project en masse. Generally, they are well funded, well organized, and limited to working in a circumscribed area because of the difficult logistics involved in moving people and equipment from one site to another. Fieldwork with large groups of researchers is most often carried out in the United States because of reduced travel costs, vehicle availability, research site accessibility, and minimal bureaucratic rules. Large overseas field projects are usually based at a single site, such as a field station, simply because of logistical difficulties. Such operations are very expensive and thus few in number.

I prefer small field crews, especially when working abroad. I find it much easier to coordinate both the trips and the research with a small group. Small crews are also more mobile than large crews and the results of their efforts are qualitatively and quantitatively different. Small groups cover large areas quickly, exploring new habitats, finding new species across many habitats, conducting comparative studies between habitats, and opening up new avenues of research. But many of the initial observations made by a small field crew can be clarified only by the long-term research of a larger group. Thus the two types of research strategies are complementary.

Those who grew up with the *Star Trek* television series of the sixties know that the primary reason the intrepid crew was sent to the stars was to seek out new life forms. In this, Gene Roddenberry's creation presaged society's fruitless and extraordinarily costly cosmic search for living creatures on other planets. Humans have gone to inordinate lengths to ascertain whether and whither life exists elsewhere, hoping to prove that earth is only one of many celestial bodies to be blessed with life.

But our own planet is teeming with life forms that have not yet been sought out. The great majority of species remain to be discovered, and their ecological roles in the world's ecosystems remain undefined. If we could move across the millennia against the current of time and trace the genetic codes of life as they mutated, multiplied, combined, and recombined over three billion years, we would find that many of our "human" characteristics are shared with all organisms. Such traits as cell structure, respiratory mechanisms, reproduction, cell metabolism, chemical transfer through cell membranes, electrolyte balance, hormonal function, and morphological symmetry are, at their broadest levels, far from unique to our species.

If we had the ability to translate the genetic codes of all life, past and present, we would discover that our chromosomes are composed of bits of the genetic information of all of the organisms that have come before us. We share genes with the simplest bacterium and our closest relative, the chimpanzee. We carry parts of the DNA of the first unicellular marine organisms, the earliest invertebrates, and even the great dinosaurs. Genetically, we are one with the creatures of our planet. We are reflections of the life that has been and harbingers of the life that is yet to be.

Our very essence as a species is thus intertwined with all other forms of life, and only a relatively few genes make us uniquely human. The genetic code forms an unbroken chain that binds us to the extinct life of the past and stretches intact into the future with a promise of eternal life, a continuous thread of being from ages past into an indistinct future that we cannot even imagine. Whatever the future holds for *Homo sapiens*, perhaps we can find solace in the fact that whether or not our species endures, life itself will persist. As long as bacteria multiply or birds fly, a part of us remains coiled within life's web.

As humans we are uniquely aware of this incongruity of life, seeing both its persistence and its ephemeral nature, its unique expression in a cornucopia of species and its similarity among all species in the genetic code of life. We are also cognizant of the fact that our species affects the existence of most of the world's other species. Sadly, we are unable to say how many species exist, and even our guesses vary widely, with published estimates of global species numbers ranging from 5 million to more than 100 million. Fewer than 2 million species have been named, and most of these are insects. Yet recent research indicates that bacteria and other unicellular organisms may be the most speciose life forms, with some even continuing to live within rocks that were buried deep within the earth more than twenty million years ago.

Each species is unique, comprised of a mix of genetic information not replicated in any other living organisms or in any other organisms that have ever lived. A new species is thus the most profoundly unknown form of life, a new combination of the life stuff itself. It may even carry information that will prove to be useful to the survival of our own species when we have learned enough about genetics to decipher the messages that are buried within the genes. We have not yet reached that level of understanding. We are still seeking the Rosetta Stone of life.

Ironically, only a small amount of money has been spent to search for new life on our planet, the only world known to harbor living organisms. From the tiny percentage of species that we have described, we derive medicines, food, clothing, and construction materials. Only three species of plants—wheat, rice, and corn—supply humanity with more than 50 percent of its food. The small percentage of the known living world that has been described, in addition to making our modern existence possible, has also inspired our finest art, poetry, literature, music, and philosophy. Nature's richness has enriched our lives.

But we are at great risk of losing the large part of the world's richness and diversity that has not yet been discovered. Research from many fields indicates that species are inexorably sliding toward extinction because of deforestation, desertification, global climate change, and other activities of the human species. If everything is lost, or will be lost in a decade or two (and this is arguable, since we do not know what is out there), nothing can be done to reverse the trend, especially given the complexities of world politics, global and national economies, and human nature. The species are doomed. If, however, as I believe, these species will disappear mainly over the span of a century or more, we still have an opportunity to conserve many of them. There is some hope that long-term programs (on the order of five to ten decades) can reverse or reduce the rate of species disappearance. Thus far, however, no one has developed a century-long conservation plan.

The outlook for the long-term persistence of the world's biota as it exists today is therefore uncertain. Nevertheless, there are biologists who travel to the most remote places on the planet to seek new species that truly are "anonymous"—without a name—to learn about species that have never before been studied, and to search for answers to broad questions about life in the field. Although no one alive today will journey into the cosmos like Captain James Tiberius Kirk and the dauntless crew of the *Enterprise*, some people have been privileged to seek out new life forms on this planet, going (sometimes even boldly) where no one has gone before. I am one of them.

The Immortal Animals

2

Systematically organized collections will always be at the
cutting edge of biological research, yet not always obviously
so, but there of necessity because if you don't name it correctly, you don't know what you are talking about.

R. Y. Edwards, 1985

Exactly where did South American chinchillas
occur before they were hunted almost to extinction? The chinchillas are
gone from most of their original range. But we know where they occurred,
when they reproduced, when they molted, how they grew, and many other
facts about their lives because there are specimens of chinchillas in the British Museum, the national museums of France, Argentina, Chile, and Peru,
and many other museums. Taken together the specimens in these collections tell us where the species was found, when it stopped being collected
(the lack of collection showing that it was on the road to extinction), and
where it exists today. They also tell us where we can hope to reintroduce endangered species so that they might have a chance of surviving in the wild,
thus avoiding the ultimate loss: extinction.

As new techniques are developed, collections become more and more
valuable. A hundred years ago the specimens had great value, but their utility was more limited than it is now. Today it is possible to take DNA from
dried museum skins for molecular, viral, and other analyses. Museum collections have become living repositories of life past and present. A good example of this is recent research on moas, extinct birds that once lived in New

Zealand. They were large, flightless birds—more than 12 feet tall—that disappeared forever in the late 1600s, killed off by habitat destruction and uncontrolled hunting. Fortunately, a very few were preserved as museum specimens and are still available today. In 2001, DNA was isolated from the bones, and scientists were able to clarify the relationships of the extinct birds with the living ratites—ostriches, rheas, cassowaries, emus, and similar species. The specimens that were fortuitously collected and preserved centuries ago permitted the use of the latest genetic technology to study the history of these birds.

Biology is the science of life, and field biology often involves the search for new life, but in order to study animals field biologists frequently have to collect them as specimens that are permanently preserved in a museum as benchmark specimens of the species. With the exception of the relatively few showy species, the only correct identification for most mammals is based on the examination of specimens preserved in a museum. These specimens provide an irrefutable record of a particular species at a particular place and time. When a species is first described, a type specimen is designated—the actual specimen on which the description and name of the species are based and which will be the benchmark for all future identifications of that species. It will have to be examined when new species are discovered in order to ascertain that they really are new and not just a variant of a known species. As additional specimens are added to museum collections they begin to form the documented record of that species over its geographic range and across time. Taken together, the specimens in a museum are the primary data on the occurrence, habitat selection, and coexistence of most of the world's mammals.

Most small mammals are difficult to identify, especially in the poorly studied parts of the world (and *most* of the world is poorly studied). One can seldom collect a living small mammal and immediately identify it with certainty. It is only after extensive research has been conducted on scientific specimens in museums that one can begin to identify the species in the field. Some field biologists might like to think that they can identify them using photographs, but it does not work that way. In some cases, one is still uncertain about whether particular species are placed in the proper genus (the next higher taxonomic category above species) or are even in the proper family (the next major category above genus).

Nor are such questions limited to small mammals. It has long been thought that there are only two species of elephants: the Asian and the African. But recent research with DNA suggests that the African forest elephant is a third species. One reason it took so long to find this out is that there are too few specimens of elephants in museum collections to permit the proper morphological research to be carried out. Had there been sufficient specimens to support such a study, measures to protect this most endangered of all elephants could have been instituted by now.

When a mammalogist collects a scientific specimen of a mammal, the collection and preservation of that specimen has no effect on the population in the wild because natural selection ensures that animals produce enough offspring for the species to survive natural calamities and the usual background rate of death. In fact, organisms produce many more offspring than the habitat can support over long periods of time. It is from these numerous offspring that the next generation's population will be selected. The winnowing force of natural selection ensures that the most fit specimens will be more likely to leave offspring to represent their genetic line in future generations. Abundance and selective death—along with mutation and other genetic mechanisms—are the molding forces of evolution. Animals die because they are programmed to die. They die from predators, starvation, disease, accidents, floods, fires, and even by becoming the victims of pet cats. Eventually, they all die, but before every member of a generation perishes, it has an opportunity to reproduce. The great majority of animals die without leaving successful offspring behind. Among some species of rodents millions, or even billions, will die each year. From this cascading river of death, the field biologist selects only a few drops for inclusion in the world's museum collections.

To put museum collections in perspective, think of all the major mammal collections in the world, the result of centuries of effort by thousands of collectors and scientists who explored the world to chronicle its mammals. Consider the British Museum of Natural History in London—a museum at the forefront of exploration and collecting for more than two centuries, the

Muséum National d'Histoire Naturelle in Paris, the Smithsonian Institution, the American Museum of Natural History, the Carnegie Museum, the Field Museum, the Academy of National Sciences of Philadelphia, and the national museums of Australia, Argentina, Brazil, Canada, China, Germany, India, Russia, South Africa, and a host of other nations. Add to the list outstanding university museums at Berkeley, Harvard, Kansas, Michigan, Oxford, Yale, and elsewhere. There are about 650 mammal collections in the world and they are in almost every country. Each holds specimens in trust for the citizens of its nation—fundamental data for wildlife managers, government agencies, public health officials, consulting firms, land use planners, foundations, schools, conservation organizations, and other groups. Collectively, they contain much of what is known about the natural history of mammals of the world.

How many scientific specimens of mammals have been gathered after several centuries of collecting? A good estimate is six million (in 1981 the most recent survey of all collections showed that there were five million specimens; I have allowed for some growth). That number includes all specimens of the roughly 4,600 mammal species of the world. More than four million are of bats and rodents, the most abundant small mammals in nature. In more than three centuries of collecting, the total number of museum specimens that has been amassed is only 0.006 percent of the number of mammals killed *each year* in the United States by pet cats. On a worldwide basis, the average numbers of mammals in museum collections gathered over centuries of effort (20,000 per year) is less than 0.0000001 percent of the *yearly* catch of mammals by pet cats. This means that only one specimen out of fifty million that would have been killed by a cat ended up in a museum. The mammalogists got one animal; the cats got 49,999,999.

You might think, "Well, small mammals are one thing and sure, there are forty million free-tailed bats in a single cave in Texas on a single day (six times the number of specimens that exist in all of the collections of the world), but what about big mammals? Collecting elephants and other large species is indefensible and harmful to the species." This is a point well taken only if one does not understand the mathematics of nature. There are probably fewer than 400 elephants in the world's collections. That is, after centuries of collecting—including the period when elephants were abundant and had just begun to be studied—only a few specimens ended up in a museum (and many of these were circus and zoo elephants). Think of that number 400 in relation to the 600,000 African elephants that are now alive, the 700,000 that were poached in the last twenty years, or the 30,000 African elephants that were poached in 1998 and 1999 alone. Museum collecting does not damage populations.

In the future, when the two—or three, if the forest elephant is a distinct species—remaining species of elephants are extinct, the only way all future generations of humans will know anything about these magnificent mammals is through museum collections. There will be the tissues, the tusks, the skin, the genes, the chromosomes, and other organs. Over the succeeding centuries, these few specimens will still be studied by the scientists. They will know where elephants occurred and why they went extinct because the specimens will tell them so.

Other mammalogists and I use many techniques to collect specimens. Larger species we hunt with a rifle or shotgun. We also catch large mammals alive in cage traps (and some may be euthanized if they are destined for a museum). Small mammals are almost always collected with traps, and there are many types of traps that cater to the different requirements of the researcher and the different habits of the mammals. The most common are similar to the mousetraps that you might use at home. They come in several sizes for large and small rodents and are designed to kill an animal instantly by breaking its back or neck without damaging parts of the body (skull, teeth) that are important in research. The animal suffers little, if at all.

Often investigators need to capture an animal alive to mark it and then release it back into nature, to collect blood or tissues for genetic analysis, or to study it in behavioral or laboratory research. We use traps that are either metal boxes or cages and that have bait to attract the animal. When it enters, a door closes and it is captured alive and unhurt. For gophers and other burrowers, different traps are required—traps that can operate underground and that will not be clogged by dirt. There are both kill and live traps for these animals, too. Sometimes leg-hold traps are used to capture carnivores and other medium-sized mammals. In these, the animal steps on a trigger that releases spring-loaded clamps (rubber coated to avoid injury) that hold the animal by a foot until the investigator arrives, at which time the animal is either euthanized or marked and released. Bats are collected in mist nets—fine nylon nets that are placed over water or across trails and that cannot be detected by the bat's sonar. The bats fly into the nets, become entangled, but remain alive and unhurt. Many (if not most) larger mammals are collected as salvage specimens that were already dead and would otherwise be left to decompose (roadkills, beached whales).

The few mammals that are prepared as museum specimens are humanely killed, sometimes with ether or carbon dioxide. More often, small mammals are killed by thoracic compression, which stops the heart quickly. In such cases an animal dies rapidly.

A Sherman live trap (3 inches × 3 inches × 10 inches) set near a stake. The trap was used in an ecological study in which animals were captured and observed, and then released. It has an electrical sensor attached to the door that indicates when an animal has entered the trap. (Photo by M. A. Mares.)

When I and other mammalogists catch an animal, we record information such as the date, the habitat, the weather, the time of capture, and the type of trap, and take standard measurements (total length, hind foot length, ear length, tail length, weight). We give each animal a unique number and tag. If more than one specimen of a particular species has been obtained, we prepare at least one individual in formalin, which is a liquid preservative that fixes the tissues and stops decomposition. After we give these formalin-preserved animals to the museum, they are transferred to ethyl alcohol for final storage. Such specimens are useful to anatomists and other biologists who do soft tissue research. The preserved specimens also include all food, intestinal contents, endoparasites, and other materials that may be of use to investigators. Tissue samples and other anatomical structures noted below are also collected from specimens preserved in fluids.

For animals that will be prepared as study skins, we follow a different procedure. We remove the skin and examine it for evidence of molting, which shows when an animal replaces its fur, a fact often related to season or nutritional status. The skin is stuffed with cotton, wire is put into the legs and tail for stiffening, and the skin is pinned into its final position and left to dry.

After the skin is prepared, work continues on the internal anatomy. We note fat content and condition of the reproductive organs. We remove and individually label the internal organs, placing samples of muscle, ear, kidney, liver, heart, testes, blood, and spleen in labeled vials and quick-freezing them in liquid nitrogen or another preserving fluid. If an animal is pregnant, we note the position of the embryos in the uterus and the embryos are measured and preserved. We examine the stomach and intestines for food content. We record all findings at each point. We may collect internal and external parasites, which means we have to prepare additional labeled vials. All of the tissues, organs, and skeletal material will always be associated with the scientifically prepared specimen, so that the correct identification of the species will be able to be verified. This will help scientists avoid publishing erroneous results. Geneticists, virologists, molecular biologists, wildlife biologists, and other scientists who require living tissues and genetic material will use these tissues in their research. All the data are entered into a computer database.

We remove marrow from the long bones and treat it in a series of complicated chemical procedures that cause the cells to divide, inflate, and stop dividing at the point where the chromosomes are most visible. We then fix the chromosomes on labeled microscope slides (blood samples are also fixed in this manner).

In many cases, we are unable to identify specimens in the field because so many species look alike. Final determination as to species is not possible until the animals are put in the museum. The skull and other bones are

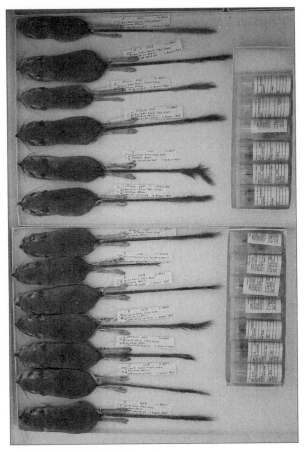

Museum study skins of kangaroo rats *(Dipodomys)* from Mexico in the Sam Noble Oklahoma Museum of Natural History in Norman. The tags contain the collector's name and number, locality, sex, museum catalogue number, date of collection, and various external measurements. The vials contain the cleaned and numbered skulls. (Photo by M. A. Mares.)

cleaned in the museum by the larvae of dermestid beetles, which feed on the dried tissue and leave the bones fairly clean. When the "bugs" are done, we rinse each individual bone with water, remove all bits of soft tissue with fine forceps, and dry the bones. We mark each bone in permanent ink with the same number that we placed on the skin tag when the animal was cataloged into the museum collection. After this we place the bones in individual boxes and store them with the dried skin, both of which have been fumigated for several weeks in a carbon dioxide enclosure or in a freezer to kill all possible insect pests.

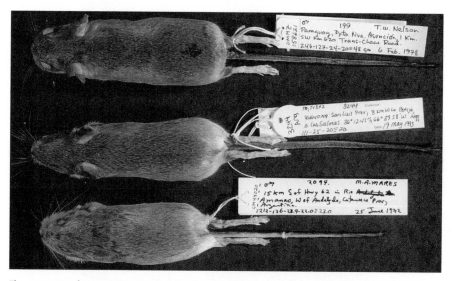

Three species of mice: Pearson's desert mouse, *Andalgalomys pearsoni* (top); Roig's desert mouse, *Andalgalomys roigi* (middle), and Olrog's desert mouse, *Andalgalomys olrogi* (bottom). The species at the top was originally placed in a different genus, *Graomys*, until the species at the bottom was discovered and placed in the proper genus, *Andalgalomys*. Now there are three species in the new genus. (Photo by M. A. Mares.)

We subject the chromosome collection on the microscope slides to an extensive series of chemical stains that differentially make the chromosomes visible against the background of the rest of the cell. We then examine the chromosome slides under a compound microscope to find those few chromosome sets that have gotten fixed, stained, and placed on the slide in a manner that permits them to be photographed and counted. We photograph them, store the images digitally, and compare them with known chromosome complements to help decide whether we are dealing with a new species or one that has already been described.

At last, we lay many specimens of several species side by side and compare

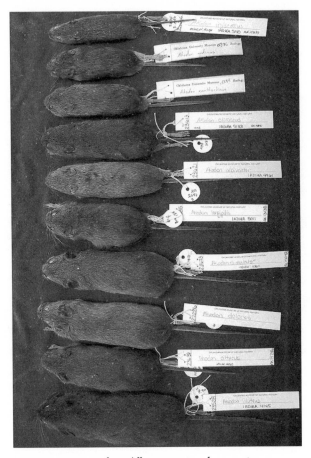

Museum specimens of ten different species of grass mice,
Akodon, laid side by side. The animals are difficult to identify outside of a museum. (Photo by M. A. Mares.)

them in detail. We also measure many features on the skin, skull, and skeleton, which we analyze with complex mathematics using a computer. Only
then can we identify them. To this point a specimen has been known only by
its field identification (for example, "*Eligmodontia*, poss. *typus*," "*Akodon*
sp.," or just "Genus?"). This process may take months, depending on the
complexity of the taxonomy of the mammals in question. If it is an animal

that is new to science, we begin the long task of preparing a scientific publication that will formally describe the species. Such work may take a year or more. If parasites or other special collections have been made, we make these available to specialists after the mammals have finally been correctly identified.

When all things are considered, each specimen takes many hundreds of hours of field, laboratory, and publication time, and no small amount of personal risk and discomfort. One reason that the museums of the world have relatively so few specimens of each species (compared with the abundance of animals in nature) is that each specimen takes an enormous amount of time, effort, and money to collect, prepare, preserve and study. Naturally enough, museums attempt to protect their collections "forever," so that researchers will always have access to the primary database of nature.

The animals that I and other mammalogists study often are extremely rare. Sometimes the only specimens that have ever existed in a museum for a particular species are the ones we collect. When we go to the great trouble to capture, euthanize, and prepare an animal, we want it to be as valuable as possible to as many investigators as possible. Scientists from throughout the world will use the materials that we prepare in all manner of research, from medicine to conservation. A part of our research involves sacrificing a few animals, just as the laboratory scientist sacrifices mice used in medical research. These very few animals, which would have died in any case in the wild, will have their skins, organs, skeletons, and living tissues studied far into the future. They will become immortal animals.

Elfin Farmers and Cactophylic Carpenters

When I get up in the morning at my desert camps, usually one of the first things I do is to look around the dead campfire to examine the myriads of tracks in the fine dust and see what rodents have been nocturnal visitors. The tracks show plainly that white-footed mice are among the most frequent guests; and more than likely the kangaroo rats have been there, too . . . It is my firm conviction that every foot, almost every square inch, of earth is hunted over each night by these harvesters.

Edmund Jaeger, *Desert Wildlife*, 1961

The abundant and fascinating life in North America's well-studied deserts first attracted me to work in arid habitats, but by working in other deserts I came to feel a special appreciation for the diversity and the wonder of desert life. During my research on other continents I learned how the evolutionary story unfolds when whole assemblages of mammals develop in areas having similar climates, but different topographies, floras, and geological and faunal histories. I learned that the way things work in one desert is not necessarily how they work in all deserts. Each organism has a unique set of genetic information, as well as its own particular story of colonization of the desert and adaptations to aridity. Mammals in different deserts may follow similar evolutionary paths in adapting to aridity, but they are also likely to have taken paths that are peculiar to them and to their individual desert. The desert story that unfolds in this book is, in many ways, a comparative tale of desert research that, like the scientific exploration of deserts, began in the United States but continued in other lands.

To understand the deserts of the world we must begin with the arid lands of North America, for that is where most desert research has been done. Although much of the United States offers bountiful habitats, fully 750,000 square miles of land are desert or semidesert, an area larger than the combined area of France and the United Kingdom. The distance from the westernmost point in the semidesert of Washington state to the southeastern-most point of the desert scrub of south Texas is 2,000 miles—greater than the distance from Stockholm to Cairo. Annual precipitation does not exceed 10 inches throughout this area.

As the waves of European colonists moved westward, they eventually came to people the arid portions of the country. In New Mexico, my home state, they encountered the Spanish settlements that had been established in the 1500s. The Spaniards had followed the Rio Grande River northward from what is now Mexico. They established the city of "Alburquerque" (the first "r" was later dropped), in north central New Mexico near the river they called the Río Bravo—Fierce River—with its green thread of riverine cottonwood forest meandering through a sparse, brown landscape. Like the Nile for the Egyptians or the Euphrates for the Babylonians, the river flowing through the New Mexico desert meant life for both the Spaniards and the Indians. With the river, crops were possible, life was tolerable, and water was available throughout the year.

Economic development and societal expansion were the driving forces for desert research at the end of the nineteenth century. Government biologists were sent forth in the late 1800s to explore the new lands that were manifestly destined for inclusion in the developing nation. As the outpost forts developed and the native peoples were conquered, pacified, or removed, westward expansion brought settlers to the parched western lands. These newly developing territories were quick to found state universities. Since there was an economic need to discover the biological resources of the newly acquired regions, natural scientists were among the first faculty members at the developing institutions. They worked to determine the best strategies for developing farms, ranches, and forests, and for utilizing wildlife and mineral resources.

Scientists in the twentieth century built upon the work of the early explorers and collectors in the Southwest. From the mid-twentieth century on, they conducted extensive research on the systematics, ecology, physiology, and behavior of both large and small mammals throughout the West, including the deserts, which extend through eleven states. The information obtained by these biologists provided the lens through which deserts would be perceived by scientists throughout the world.

On the popular front, western movies were a uniquely American cultural influence that affected how the public viewed deserts. Westerns of the 1930s to the 1950s—the most popular film genre of the century—were invariably set in deserts. Classics such as the John Ford/John Wayne westerns projected an indelible cinematic image of the desert in the United States and throughout the world. Eventually, even in the popular mind, the deserts of North America came to be viewed as exemplars of deserts everywhere, and information on desert organisms in the arid United States became the paradigmatic data for organisms in all deserts.

The most northern desert in the United States is the Great Basin, which lies mainly in Nevada and Utah. It is a cold desert, receiving about half of its yearly precipitation as snow. Among its most common plants are shrubs of the goosefoot family (Chenopodiaceae), such as saltbush *(Atriplex)*, and desert-adapted shrubs such as sagebrush *(Artemisia)* of the sunflower family (Asteraceae). The mix of grasses and sagebrush, as well as the extensive areas with salty soils that support saltbush, help define this desert. Unlike the plants in many hot deserts, which either lose their leaves or have their leaves shrivel during the most arid seasons, saltbush and sagebrush remain luxuriant all year.

Saltbush, species of which are found throughout the world's arid areas, illustrates the remarkable adaptations of plants for desert survival, especially in its interactions with mammals. The plant's fluids contain high levels of sodium chloride (salt) that permit the plant to retain water in its tissues in spite of growing in salty soil. As rain falls, it percolates into the soil, where it dissolves the salts. This salty water eventually moves toward the surface as evap-

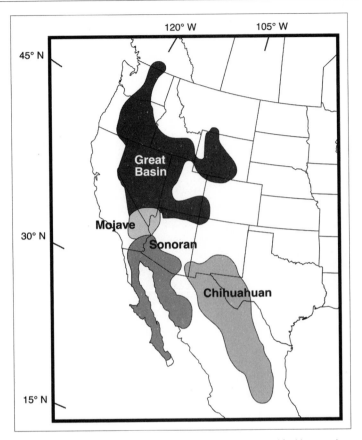

The deserts of North America. (Map by Cartesia MapArt, modified by Patrick Fisher.)

oration occurs, leaving the salts deposited on the soil surface. In some areas salt flats may develop into white, plant-free deposits of pure salt that cover hundreds of square miles.

The challenge to plants inhabiting salty soils is that water moves from low to high salt concentrations. In salty areas most plants will lose water from their tissues to the soil, but saltbush has found a way to concentrate salts in its tissues to levels that are so high that they permit the plant to draw moisture from the saltiest substrate. Saltbush leaves are saltier than sea water. As the saltbush plants move salts from the soil into the plant, and then

The Great Basin Desert in Idaho, with sagebrush and saltbush. This desert strongly resembles the Patagonian Desert of Argentina. (Photo by M. A. Mares.)

through the plant and out the leaves, salt crystals are deposited in special cells on the leaf surface, making the leaves appear gray or silver rather than green. The salts also form an effective barrier against herbivores, protecting the plants against organisms that might attempt to forage on the luxuriant foliage.

High salt concentrations are toxic to most organisms. Think of a human adrift at sea, like Coleridge's Ancient Mariner: "Water, water, everywhere, Nor any drop to drink." The reason we should not drink sea water is that it causes our body fluids to increase their salt concentration, which impels our system to rid itself of this toxin (at high concentrations) by flushing the salts from the body. The only water available to do this is in the tissues. Thus if sea water is used to slake thirst, more water is lost ridding the body of salt than is obtained from the sea water. Saltbush is tempting to herbivores because it contains not only nutrients but also water, but its salty tissues make it a forbidden food for most. Few animals have solved the puzzle of how to eat its leaves and survive.

Life finds a way, however, and several species of mammals have managed to overcome the defenses of these halophytic (salt-loving) plants. One of these is an unusual member of a remarkable group of rodents that inhabits North American deserts, the kangaroo rat, a relatively large bipedal rodent. Kangaroo rats are seldom seen because, like most desert mammals, they are active only at night. They belong to a family (Heteromyidae) that also includes pocket mice (small, quadrupedal, seed-eating rodents) and kangaroo mice (mouse-sized, bipedal rodents that hop on their hind legs).Pocket mice are tiny, weighing 20 grams or less—the silky pocket mouse weighs only 6 grams, the same as a small lollipop. A few kangaroo rats, such as the desert kangaroo rat, may reach 150 grams or more (5 ounces), and are impressive members of the desert's small mammal fauna, but most kangaroo rats weigh about 50 grams (about 2 ounces).

A desert kangaroo rat, *Dipodomys desert.* (Photo by T. L. Best; Mammal Slide Library, American Society of Mammalogists.)

The bipedal kangaroo rats are paragons of adaptation to aridity and, their small size notwithstanding, are extraordinary in many ways. They are considered to be among the most highly specialized desert mammals in the world. Consider the challenges they face. If you were to find yourself isolated in prime kangaroo rat habitat for more than a few days you would die of heat, thirst, exposure, or a combination thereof. Not so these little creatures. In a desert, vegetation is scarce, so these plant eaters must find a way to survive, even though plants are literally few and far between. They specialize on the tiny seeds produced by desert plants—seeds so minuscule that they are practically impossible to see on the desert floor.

To gather enough of these seeds to survive, the kangaroo rat uses its forefeet to winnow through the desert soil, separating out the seeds from the sand. The rat does this so rapidly that it is difficult to see the front feet move and impossible to see the seeds being gathered—up to 60 seeds per second. Special photography shows the seeds being carefully and rapidly sifted from the sand by the forefeet. The head is held near the soil surface and on either side of the mouth there is a fur-lined external cheek pouch into which the seeds are tossed. After filling the pouches the rat buries the seeds in storage chambers in its complex burrows or in shallow surface caches near burrow entrances.

Seeds are an ideal desert food, being high in energy and remaining viable for years or even decades until the next rain. The accumulation of seeds in the soil (called the seed bank) is what leads to the spectacular desert blooms of annual plants that follow heavy rains. The desert is suddenly and almost miraculously filled with colorful plants that seem to spring to life overnight. In essence, desert annuals spend most of their life as seeds waiting for the infrequent rains to stimulate germination, growth, and flowering, all of which may take only a few weeks. Their seeds may fall into cracks in the soil, may be covered by blowing sand, may fall into spider and lizard burrows, or may otherwise become buried. Kangaroo rats and pocket mice have an acute sense of smell that detects the odor of buried seeds. The seeds might have been buried for years by the time the rodent comes searching for them, but it rapidly digs them up, fills its pouches, and hops off to bury them.

The seeds of annual plants are the long-lasting packets of high-energy food that are the mainstay of the diet of most kangaroo rats. But these seeds also serve other functions. The rat seals the burrow openings during the day to keep out the dry desert air. It breathes the more humid air of the enclosed burrow and the moist air that is expelled from its lungs increases the humidity of the burrow even more. The stored dry seeds absorb this moisture, and so when the rat eats the seeds it recaptures some of the water that it lost in the exhaled air. The rat thus loses less water by eating the stored seeds than if it had eaten them as they were gathered.

Storing seeds underground is not easy. One problem is that the seeds will mold if they become too moist, so the kangaroo rats, little farmers that they are, move the seeds around to drier parts of the burrow, keeping them from becoming inedible. Exactly how they accomplish this, and whether they plug and unplug parts of the burrow to adjust airflow and humidity is not clear, but they control the moisture level of the seeds by moving them around and drying them when necessary. They do this year-round with millions of seeds gathered from hundreds of plants at different times of the year under varying weather conditions.

Some investigators have suggested that kangaroo rats store seeds in such a way that they cause medicinal fungi to grow, thus acting not only as diminutive farmers, but also as tiny pharmacists. Molds and fungi do grow on the seeds and, in principle, one might expect that such an important attribute as developing an antibacterial fungus would be grist for the mill of natural selection. Animals that cared for their seeds in the proper manner to produce medicines would not suffer as many diseases as animals that did not grow the appropriate molds and fungi. So far this has not been proven, but nature will continually surprise us if we look closely enough.

Hunting for tiny seeds at night in the desert is challenging, especially since kangaroo rats avoid foraging on moonlit nights, when predators can easily see them. The desert is a dangerous place. Not only is it not lifeless, it is literally crawling with life, including many species that eat kangaroo rats, among them rattlesnakes, bullsnakes, coachwhips, king snakes, owls, coyotes, bobcats, and foxes. How do the kangaroo rats find the tiny seeds in the dark?

In addition to being able to detect the odor of seeds buried 8 inches below the surface, a kangaroo rat has dozens of extremely long, tactile whiskers. These act like the cane of a blind person and help the animal feel its way along as it moves through the dark night. And the large eyes set high on top of the head allow the rat to see at very low light levels and to see over a much larger arc of vision than humans. It is thus able to watch for deadly predators while it gathers the thousands of seeds it needs to collect each night to survive.

In addition to the specially adapted eyes, the bony chambers enclosing the inner ear are inflated to an extraordinary degree, magnifying the slightest sounds that ripple through the desert air. With more than a third of its skull devoted to sound-amplifying chambers, a kangaroo rat can hear even the soft wingbeats of owls in the dark velvet night and the scraping of a snake's scales on the dry desert sand, sounds that we cannot perceive.

Kangaroo rats also have highly honed reflexes, and will leap high into the air when they sense any sudden movement or sound. Even the blindingly fast strike of a rattlesnake is often too slow to connect with the agile kangaroo rat. Kangaroo rats bound across the desert while zigzagging unpredictably from side to side. The long tail with a tuft of long hairs on the end counterbalances their weight. The full function of the tufted tail is not clear. The tuft may permit greater maneuverability, its position at the end of a long tail helping to steer the animal even while it is in mid-air. Presumably, pulling the tail through the air causes drag, which resists and opposes the rat's movements. The tuft may also attract the strike of a predator. In some species the tuft is a startling black and white color, like a flag. If a predator grabs the tuft, it will break away, permitting the rodent to escape. In some species, if the tuft is lost to a predator it redevelops on a shortened tail, which indicates that its function is important enough for it to regenerate if it is lost.

Almost all kangaroo rats and their relatives are specialized to live on seeds. Like all rodents they have an upper and lower pair of incisors (the first two teeth in the jaw). In most kangaroo rats the incisors come to a point, ideal for concentrating the biting force and breaking the tough coats of desert seeds. But there is an unusual kangaroo rat whose incisors are shaped very differently from those of other kangaroo rats—they are squared off, much like a chisel. In fact, this species, an inhabitant of the Great Basin Desert, is known as the chisel-toothed kangaroo rat. Its lower incisors, when viewed from the front, have very flat, broad tips, and when viewed from the side, taper to a thin blade exactly like a wood chisel. The shape of the teeth is related to the unusual diet of this species.

The chisel-toothed kangaroo rat *(Dipodomys microps)* lives in areas that support saltbush plants. Unlike its seed-eating relatives, this rat survives mainly on a diet of saltbush leaves. Since few mammals eat saltbush because of its extremely salty tissues, the obvious question is how did the rat manage to eat the salt contained in the plant? The chisel-shaped incisors serve an un-expected function. The animal takes the gray-green, salt-covered leaves of the saltbush in its forefeet and very carefully and deliberately scrapes each leaf against its lower incisors to strip away the salt-filled layer of cells that cov-ers the leaves. The salty tissues fall to the ground, leaving behind the green and succulent leaf tissues.

This kangaroo rat, unlike its brethren, does not store seeds, for saltbush bears green leaves throughout the year. When it developed the ability to feed on this plant, the chisel-toothed kangaroo rat no longer had to compete with smaller pocket mice for little seeds and with similar-sized kangaroo rats for larger seeds. It became a specialist associated with the saltbush plant, and it is the only rodent in North America that was able to overcome the sig-nificant defenses of this desert plant.

Of all of the deserts of North America, it is the Sonoran Desert, lying south of the Great Basin and the Mojave Desert, that most readily comes to mind when we think of a desert. The Sonoran Desert is a hot desert that receives all of its precipitation in the form of rain. The Sonoran is the lushest North American desert, with the tall, many-armed, saguaro cactus *(Carnegiea gigantea)* as its most characteristic plant. Indeed, some desert scientists do not classify this desert as a true desert because it receives as much as 10 inches of rain or more every year; the "classic" definition of a true desert (hyperarid area) allows for only a few inches of precipitation a year. (The Great Basin and Chihuahuan deserts also fall into this moist desert cate-gory.) The Sonoran is home to paloverde trees *(Cercidium)*, yellow-green or bluish-green trees that carry on photosynthesis within their green bark, put forth yellow flowers in the spring, and have tiny leaves that minimize water loss. The desert also supports the odd ocotillo plant *(Fouquieria splendens),* which resembles a bundle of thin dead sticks for most of the year. It produces thick green leaves when rains fall, only to lose them again as the drought returns, until the next rain brings forth another crop of leaves. Each year in the spring the ocotillo produces bunches of carmine flowers that attract hummingbirds and insects, flowers that often seem to be balanced on the ends of the dead, spiny branches.

Sonoran Desert vegetation near Tucson, Arizona, with several saguaro cacti *(Carnegiea gigantea).* (Photo by M. A. Mares.)

In some areas this rich desert supports a forest with trees of mesquite *(Prosopis),* paloverde, ironwood *(Olneya),* cottonwood *(Populus),* acacia *(Acacia),* and saguaros. Between the large cacti and the trees are legions of other cacti, including many species of cholla, the multiple-armed and mightily thorned cacti 3 to 6 feet tall that drop their segments to form new cacti. They have names like staghorn cholla, teddy bear cholla, cane cholla, pencil cholla, and jumping cholla, names that describe how they look or what they do. Jumping cholla (*Opuntia fulgida*), for instance, is a species that has a propensity to break off segments easily, which then bounce crazily off the long, stiff, springlike thorns after they fall to the ground.

When my son Gabriel was six years old and learning the ways of the desert while accompanying me on field trips, I showed him how to identify and avoid jumping cholla. Shortly after his first cholla lesson he proceeded to obtain a more direct education from the cholla itself. Being cautious by nature, Gabriel found a long stick and decided to experiment with a large jumping cholla by poking it until its segments broke off. He believed his dad's story about the dangers of the cactus, but thought that he could out-smart the plant. The length of the stick imparted an illusion of safety as he jabbed at the cactus. Within seconds a large segment broke off and bounced rapidly downhill toward Gabriel's spindly and exposed legs. I heard him scream and ran to help. The cholla had buried its thorns deep into his shin, perhaps reaching the bone. Realizing that he had been prodding the jump-ing cholla shortly after I had warned him not to, Gabriel did not cry out after his initial scream, although tears rolled down his face. The cholla's lesson was far more effective than mine.

Cholla is such a dominant plant in the Sonoran Desert that several species of animals have found a way to use it as an important resource. One of these is a rodent called the white-throated woodrat (*Neotoma albigula*), an animal that not only feeds on cholla, but is a master carpenter that uses the spiny cactus as a building material to provide protection from predators. This woodrat (also called a packrat) weighs about 200 grams (or 7 ounces) and eats cactus.

Cactus might seem like a good plant to eat if you can get past the spines, since the succulent branches contain both water and nutrients. But the cholla has more complex defenses than just its spines. Cholla produces ox-alic acid, which is an organic acid that is used as a fungicide and as a bleach in woodworking and leatherworking. Oxalic acid is highly toxic to mammals. Humans would die if they ate oxalic acid, developing such terrible symptoms as tremors, convulsions, abdominal pain, vomiting, weak pulse, low blood pressure, and shock before succumbing. As little as half an ounce can kill a person within minutes.

Oxalic acid content notwithstanding, woodrats have the ability to eat as much cactus as they like. Half their diet consists of cactus. They thrive on it. The physiological mechanisms involved are not clear, but the fact that they feed on cacti is well known and was observed long ago by many early naturalists. Woodrats can also climb into and out of the thorny cholla, a feat few other animals can accomplish. When a woodrat is pierced by a spine, which happens occasionally, the animal calmly removes it. Other animals would panic with cactus spines imbedded in their flesh.

A white-throated woodrat *(Neotoma albigula)* climbing in a cholla cactus *(Opuntia)* in the Sonoran Desert near Tucson, Arizona. (Photo by M. A. Mares.)

The woodrat carries the cactus segments from the cactus patch to its home, which is constructed of sticks, twigs, and pieces of cholla. The animals also scatter cacti around the nest as additional protection from intruders. The more cholla that grows in an area, the more woodrats the area supports. Woodrat homes constructed of cactus are almost impenetrable to larger animals. But the stick nests offer refuge to many species, including vertebrates, such as the zebra-tailed lizard, tree frogs, voles, skunks, rabbits, the insectivorous desert shrew (the smallest desert mammal, at one-seventh of an ounce, an insect-sized mammal), and a wide variety of invertebrates, including spiders and scorpions.

Like the saltbush-eating kangaroo rat of the Great Basin desert, the several species of cactus-eating woodrats of the hot deserts of the United States (at least one species occurs in each of the Sonoran, Mojave, and Chihuahuan deserts) are marvels of desert adaptation. Although they are not able to live without free water (drinking water or moisture in food), they are able to obtain water from cacti, even during the driest periods. Woodrats are the most cactus-loving rodents in the world.

Woodrats defecate and urinate in well-defined areas near their nests, and over the millennia, these dung piles have hardened. They contain bits of twigs, leaves, seeds, and other material that either was eaten by the animals or got trapped in the dung. When the nests were located in caves, under overhanging boulders, or in other protected places, the hardened piles were largely immune to decomposition, and paleobotanists have found that they provide an irreplaceable record of both climate and vegetation changes over millennia. The dung piles, called middens, can be sectioned and radiocarbon-dated to provide a continuous record of habitats over a broad area for tens of thousands of years. They are one of the most effective tools in reconstucting ancient climates.

The hottest desert in the United States, and perhaps the most challenging to life, is the Mojave Desert of California and Arizona, within which lies Death Valley, the lowest point in the Western Hemisphere at 282 feet below sea level. The Mojave, the smallest U.S. desert, lies south of the Great Basin and west of the Sonoran, and is the only desert completely contained within the

United States. It is also the only U.S. desert that is recognized worldwide as a true desert. Not only does it receive very little rain—only 1–2 inches per year—but it also experiences extremely high temperatures.

Death Valley in the Mojave Desert. (Photo by M. A. Mares.)

The Mojave, in the popular mind at least, is an arid, almost lifeless desert with multicolored, starkly eroded mountains. Each year some Mojave Desert town records the highest temperature of the summer. Indeed, the highest temperature ever recorded in the United States, and the second highest ever recorded in the world, 134°F, was measured in the Mojave Desert. Ground temperatures of 190°F have been measured, only 12° below the temperature of boiling water.

Because of its aridity, the Mojave supports only sparse, low shrubs, such as creosote bush *(Larrea tridentata)*, and the Joshua Tree *(Yucca brevifolia)*, a yucca that may grow more than 15 feet high. Cacti are present, but are neither diverse nor abundant. It is too dry for cacti to flourish. The Mojave is so arid that few perennial plants grow there, but it is well known for its diverse flora of annual plants (600 species of plants are known from Death Valley alone, a small part of the greater Mojave Desert). As befits an exceptionally arid and isolated desert, fully 25 percent of the plant species in this desert are endemics, species found nowhere else on earth. Endemic fish have also developed in the Mojave, including the desert pupfish *(Cyprinodon diabolis)*, which occurs in a single sinkhole and has what is probably the smallest geographic distribution of any vertebrate (just over 200 square feet). Even fish that live in deserts inhabit extreme environments. The water temperature of pupfish pools may reach 113°F and the salt concentration may be four times that of sea water.

The mammal fauna of the Mojave in many ways parallels the fauna of the other deserts of North America. Among the larger mammals are the desert bighorn *(Ovis canadensis)*, the puma *(Puma concolor)*, the coyote *(Canis latrans)*, the kit fox *(Vulpes velox)*, the mule deer *(Odocoileus hemionus)*, and the bobcat *(Lynx rufus)*. Small mammals are also diverse. The heteromyid rodents (kangaroo rats and pocket mice, members of the family Heteromyidae) include an array of kangaroo rats—the desert kangaroo rat *(Dipodomys deserti)*, Merriam's kangaroo rat *(Dipodomys merriami)*, the chisel-toothed kangaroo rat *(Dipodomys microps)*, the Panamint kangaroo rat *(Dipodomys panamintinus)*—pocket mice—the little pocket mouse *(Perognathus longimembris)*, the desert pocket mouse *(Chaetodipus penicillatus)*, the spiny pocket mouse *(Chaetodipus spinatus)*, and the longtail pocket mouse *(Chaetodipus formosus)*—and the pale kangaroo mouse *(Microdipodops pallidus)*. There are also several rodents of the field mouse group (sigmodontine rodents), such as the western harvest mouse *(Reithrodontomys megalotis)*, the cactus mouse *(Peromyscus eremicus)*, the canyon mouse *(Peromyscus crinitus)*, the deer mouse *(Peromyscus maniculatus)*, the brush mouse *(Peromyscus boylii)*, the piñon mouse *(Peromyscus truei)*, the southern grasshopper mouse *(Onychomys torridus)*, and the desert woodrat *(Neotoma lepida)*.

Although the Mojave Desert has a rich fauna of rodents, it is less diverse than the rodent fauna in either the Sonoran or the Chihuahuan desert. An especially interesting small mammal that occurs in the Mojave is the white-tailed antelope ground squirrel (*Ammospermophilus leucurus*). Almost all desert rodents are nocturnal, spending most of their time in their burrows, especially the hot desert days. Ground squirrels, however, are an exception. They are diurnal and can be seen scurrying around the most arid habitats during the hottest days.

Each of North America's deserts supports from one to three species of ground squirrels, which are placed in two different genera (*Ammospermophilus* and *Spermophilus*). The former contains three species (one each in the Mojave, Chihuahuan, and Sonoran deserts), and the latter contains six or more arid land species (the taxonomy of the group has not been adequately delineated), including at least one in each of the four major U.S. deserts. The white-tailed antelope ground squirrel of the Mojave is one of the most desert-adapted squirrels in the United States. It is about 6 inches long and weighs about 4 ounces. Its common name derives from its habit of holding its rather bushy tail, which is white below, laid forward on its back as it runs, giving it the appearance of the white rump patch of a pronghorn. The tail is not swished from side to side, but is tapped against the back, making a small thumping noise.

Ground squirrels master the desert differently than do the kangaroo rats and their relatives or the field mice and woodrats. Kangaroo rats survive in the desert indefinitely without having access to free water. They do this by minimizing water loss at all levels and by using metabolic water (the water that results from the breakdown of the food contained in seeds) to meet their water needs. They also venture out of their burrows only in the cool desert evening. Field mice may be quite adept at living with minimal water, although most do not survive for more than a couple of weeks without access to moist foods or water. They spend long periods of time in their burrows, eat a wide array of food, and forage only at night. Woodrats, those cactus-eating rodents, can survive only for a week or two without free water and must obtain significant amounts of moisture from their succulent food. They are generally nocturnal, but are sometimes active in the late afternoon.

The Texas antelope ground squirrel *(Ammospermophilus interpres)* of the Chihuahuan Desert. (Photo by D. J. Hafner; Mammal Slide Library, American Society of Mammalogists.)

Ground squirrels are active during the day, although they mostly avoid excessive temperatures. They can survive for more than a month without access to free water, which is not as impressive as the record of the kangaroo rats, perhaps, but shows an ability to withstand desiccation that exceeds that of almost all other small mammals that inhabit the deserts of the United States. The ground squirrel's kidneys are specialized to produce highly concentrated urine, a water-saving adaptation. Although the urine is not as concentrated as that of kangaroo rats, it is far more concentrated than that of woodrats. Unlike their nocturnal neighbors, the squirrels are capable of allowing their body temperature to rise to more than 104°F for short periods of time, a body temperature that would cause a human to become lethargic and largely incapacitated. Thus, although they forage during the heat of the day, they spend only short periods exposed to extreme heat. They frequently take shelter in their cool burrows or in the shade of a tree or shrub, dissipating the heat stored in their bodies by radiating it away to the cooler microclimate, and and they do not lose any water in the process because rodents do not sweat.

The food of ground squirrels consists of both plant and animal matter, including seeds, leaves, invertebrates, and even vertebrates. The squirrels must search over rather large areas for their food and may range over 15 acres of sparse desert habitat as they forage. They climb readily into shrubs or trees in search of food. Their catholic diet supplies both their nutritional needs and their water requirements. Many squirrels in the genus *Spermophilus* are able to hibernate during the winter or estivate during the summer. These periods of reduced metabolic rate while the animals sleep in their burrows reduce the need for water and nutrients during parts of the year. The antelope ground squirrel, however, cannot hibernate, at least in the Mojave Desert. These squirrels are active all year and must find food in the sparse desert during all seasons if they are to survive until the next period of abundance when rains once again come to the arid desert mesas and valleys they inhabit. Then the animals can mate, which they do in the Mojave over a frenzied two-week period. They then return to their solitary existence and the female produces her litter of about nine young. Ground squirrels may live for several years, so additional litters may be produced in succeeding years.

The last of the four North American deserts is the Chihuahuan Desert, which in the United States lies mainly in southern New Mexico and western Texas. This desert extends into central and eastern Mexico and is best known for its immense expanses of creosote bush, which in Mexico is called *gobernadora*, or governor, because of its dominance of enormous tracts of desert. In addition to creosote bush, the desert supports grasses, yuccas, and agaves. Three plants in the Chihuahuan Desert vie for the title of the most characteristic plant: the soaptree yucca *(Yucca elata)*, the lechuguilla *(Agave lechuguilla)*, and Parry's century plant *(Agave parryi)*. Each has a beautiful white or yellow bloom supported at the end of a single stalk and grows from a dense cluster of leaves at its base. When these plants are in bloom the desert is dotted with thousands of large, pale bouquets held aloft on narrow stalks and starkly profiled against the azure sky. The Chihuahuan Desert is rich in both fauna and flora, but only the northern portions of this desert have received significant attention from researchers over an extended period of time. Among its small mammals is a most unusual species.

The northern grasshopper mouse *(Onychomys leucogaster)* is a small rodent that is a member of a genus that is unique among the desert rodents of
the world. The animal weighs less than 2 ounces and measures only about 4
inches in length, but this stocky mouse with the short tail is a fierce predator
that prowls the desert in search of both invertebrate and vertebrate prey. As
its common name indicates, it feeds on grasshoppers, scorpions (it is also
sometimes called the scorpion mouse), beetles, moths, and other invertebrates. But it also feeds on other rodents, which the muscular grasshopper
mouse is quick to attack. These tiny mice have been known to kill and eat
deer mice, pocket mice, and voles, and in captivity they have killed animals
as large as kangaroo rats and cotton rats.

A northern grasshopper mouse *(Onychomys leucogaster).* (Photo
by R. B. Forbes; Mammal Slide Library, American Society of
Mammalogists.)

The grasshopper mouse is extremely aggressive, killing its victim with a
swift bite to the base of the skull. During certain parts of the year the mice
will feed on seeds and green plant matter, but they are always ready to con-
sume prey when they encounter them. Grasshopper mice take readily to water, swimming across streams or pools to escape predators or to seek prey.
They maintain large territories in the sparse desert shrub and generally exist
at very low population densities.

At night the grasshopper mouse emits high-pitched howls, which it produces by raising its snout to the sky like a Lilliputian wolf. This habit led to
yet another sobriquet: the wolf mouse. Grasshopper mice and at least two

other species in the same genus occur in all deserts of the United States. In all of the deserts of the world they are the only rodents that became carnivorous. The factors that led to the evolution of such an odd set of species are not clear, but one reason may be that in other deserts much older groups fill the tiny carnivore niche. In Old World deserts insectivores such as hedgehogs are predatory. In the South American and Australian deserts some marsupials feed on rodents or their young. Only in North America is there a small rodent that roams the desert on the darkest nights seeking other rodents as prey, and only in North America does that small predator rear up on its hind legs and howl like a wolf.

The fauna of small mammals of the North American deserts may include more than a dozen species in any particular patch of habitat, but across the expanse of arid habitats of North America mammals occur in repeatable subsets of species. Most common are the kangaroo rats and pocket mice, but there is always a sprinkling of field mice or woodrats. Some field mice, such as the canyon mouse (*Peromyscus crinitus*) and the cactus mouse (*Peromyscus eremicus*), are highly adapted to live with minimal amounts of water and may be found in the most arid habitats. Similarly, the grasshopper mouse will occur wherever there are sufficient numbers of vertebrate and invertebrate prey to support the miniature predator. Diurnal ground squirrels are found in most areas, their presence readily noted as they scamper from bush to bush across the sun-splashed desert floor. Less commonly, gophers are found in the desert. As they construct their extensive burrows, the diggings of these underground denizens are deposited as a series of mounds on the bare desert floor that are the only sign of the presence of the animals. Finally, the rare desert shrew may be found, especially if woodrat nests are present to provide shelter for these tiny animals.

This is a brief overview of the small desert mammals of North America. A few other species may appear sporadically, such as cotton rats (*Sigmodon*), voles (*Lemmiscus*), kangaroo mice (*Microdipodops*), cottontails (*Sylvilagus* and *Brachylagus*), and jackrabbits *(Lepus)*. Jackrabbits occur in most habitats; cottontails are less widespread in arid areas. Some mammals live in the desert as part of larger geographic ranges that extend mainly into mountains, forests, or grasslands—the deer mouse, *Peromyscus maniculatus*, is such a species. Others are localized in distribution, being found only in small habitat patches in particular deserts; the sagebrush vole, *Lemmiscus curtatus*, for example, is a colonial rodent limited to areas of sagebrush and other green desert plants in the Great Basin Desert. In the same areas where the sagebrush vole occurs, the diminutive pygmy rabbit, *Brachylagus idahoensis*, is also found. This is the smallest North American rabbit, weighing 12 ounces, about the same as three hamburger patties, and one of the most unusual. Big sagebrush, *Artemisia tridentata*, comprises 99 percent of its winter diet and it is the only rabbit to forage on the plant. Clearly, the species has been associated with sagebrush for an extremely long period of time, long enough for both a distinct genus and species of rabbit to have evolved.

The bipedal kangaroo rats were the animals that first captured the imagination of desert researchers because of their many specialized adaptations to desert life. The fact that some other rodents in other deserts looked like kangaroo rats led to the idea that all small mammals in deserts tended to evolve toward a particular endpoint of desert specialization. Unfortunately, few scientists have done comparative desert research beyond the level of physiological adaptations. Eventually, I would come to study the ecology and evolution of the small desert mammals of the world.

Darkness and the Cave of the Jaguar

Life there is so challenging that you feel you've done something really fine—really rewarding—just by surviving from
one day to the next.

Craig Packer, *Into Africa*, 1994

I enrolled as an undergraduate biology major at
the University of New Mexico in Albuquerque in 1963, intending to pursue a
career in medicine. As I continued to take courses at the university, my inter-
ests shifted to zoology. In fall 1965 I completed a course in mammalogy and
took my first foreign field trip to Mexico with the instructor and his graduate
students during winter break in early 1966. The object of the trip was to collect mammals so that the university's museum would have a representative
collection of specimens that could be used in preparing books and mono-
graphs on the poorly known fauna of Mexico. We were in arid Sonora, Mexico, near the town of Carbó, looking for a cave called Cueva del Tigre, the
Cave of the Jaguar, from which, we had heard, "many bats have been taken."
We searched for the cave, which had to be nearby, but our directions were
sketchy and we had no luck finding it. I spoke with local people and one of
them showed us which small rocky hill in the distance contained Cueva del
Tigre. He also pointed to a nearby hill and said that it held a cave that was
much larger than Cueva del Tigre and had "muchos ratones voladores"—
many bats, which are called flying rats in Mexico. I began honking the
truck's horn to gather the crew and let them know that we had found a cave.

As we approached the entrance, the smell of guano and ammonia was over-powering.

Our group split up as we entered and several headed up a chimney. A fellow undergraduate and I followed a tunnel deep into the cave. In the hot, black tunnel, the temperature was almost 90°F, and our small flashlights offered only a hesitant challenge to the overpowering darkness. With each step the smell of ammonia became stronger, engulfing us and burning our eyes, mouths, and noses. Suddenly we came out of the tunnel into a large chamber filled with the characteristic clicks of bats swirling invisibly above us.

We estimated perhaps 10,000 bats fluttering around in the blackness, but who could tell. With one swing of a hand net we collected two of them. It was a species we had never seen before, so we hurriedly carried them back up the tunnel to the professor for identification. The bats were naked-backed bats (*Pteronotus davyi*, a member of the leaf-chinned bat family, Mormoopidae), a species even our instructor, an authority on bats, had never seen alive. This species is an unusual bat in that its wings are united above the back, giving it a leathery appearance. But if one lifts the connected wings, the back underneath is fully furred.

The crew raced down the tunnel and back into the main room of the cave as the fluttering bats filled the blackness, the leathery slap of their wings brushing us lightly as thousands of them swirled around us. We collected more than forty bats that morning, a small sample of the many thousands that inhabited the cave. It took until well after dark to prepare the museum study specimens that we had collected in less than an hour. I would learn that twelve hours of hard work preparing skulls, skins, and skeletons for a few hours collecting effort was the norm in field research.

January 30, 1966, began like so many days in the field. Nothing in the cold morning chill would mark it as a day different from any other, but I would find that it changed my life forever. We persuaded the man who had told us about the cave the day before to lead us to Cueva del Tigre, our original destination. The entrance was over an enormously long and aged log protruding from the end of a dark tunnel hewn into the rock. The log extended into the blackness of the cave well beyond the reach of our feeble flashlight beams.

Brazilian free-tailed bats *(Tadarida brasiliensis)* dropping from the bat-encrusted ceiling of a cave. (Photo by M. A. Mares.)

Indeed, whether we shone the light down into the void traversed by the log, up toward the roof, or along the length of the log itself, the frail beams were swallowed by the darkness.

Common sense told us that the log had to be supported at both ends, and the Mexican who guided us to the cave told us that guano miners had carried tons of guano across this very log fifty years earlier, so we decided to cross the log and enter the cave. Each of us made the perilous crossing without falling into the void, stifling the urge to gag at the stench of aged guano. We found no live bats, but two of us worked our way up a steep, narrow chimney that was knee-deep in powdery guano. Dried guano poured over me as I plowed through the ancient droppings. My hair was covered with it. My clothes were coated with it. I could feel the guano dust in my eyes, nose, mouth, and lungs.

We saw a bat hanging motionless and quickly sneaked up on it before it could fly away. It was not going anywhere, however, for it was a mummified leaf-nosed bat (Macrotus waterhousii) that had been dead for anywhere from a year to a decade. Its skin had the consistency of ancient parchment. It was the first member of the family Phyllostomidae (New World fruit bats) that either of us had ever seen.

Later I explored alone another long tunnel in the cave and found an enormous room. I could no longer hear the sounds my companions made as I climbed up a massive rock slab that protruded 50 feet into the inky dark like the prow of a ship. I shinnied to the tip of the rock, which was only a foot wide. When I shone my headlight around, the vast cavern was filled with the ghostly flapping of bat wings and thousands of high-pitched chirps. I marveled at the sight and sound, until my headlight suddenly failed.

Now I was perched on a narrow rock in a black void far above the boulder-strewn floor with thousands of bats fluttering about me. The darkness was becoming palpable. I stifled the panic that was rising in me. I called out to my friends, but they were far away behind tons of rock and could not hear me. They did not have any idea where I had gone in the massive and complex cave, with its numerous rooms, tunnels, chasms, and chimneys. As I straddled the rock in the hot, black room, the cave seemed to tilt. My senses were reacting to the absolute blackness, the tens of thousands of fluttering wings, and the hot ammonia-filled air.

Eventually, I decided that it could be hours before anyone found me. My precarious perch would only become more uncertain as time passed and my sense of balance continued to deteriorate. Getting out of this pickle was clearly going to be up to me. Working only by feel, I disassembled the flashlight, praying that I would not drop the bulb or the spring, or lose my tenuous hold on the rock and plunge into the black emptiness. Slowly, I reassembled the light after adjusting the bulb and switch in the absolute darkness. Holding on to the light with both hands, I pushed the switch and was rewarded with a flood of glorious illumination. My stone perch was still rock solid and the way out was clear.

The next day my companions and I explored an abandoned silver mine. This was a dangerous place, with creaky shoring and many collapsed areas, but we went in anyway because we had heard that the mine was a home for vampire bats. Vampires, which feed on blood, carry rabies and most Latin American countries have vampire control programs that attempt to poison or otherwise eliminate these bats, which are a scourge to both domestic animals and people. Now we would try to find this species in the dark mine. The mine was filled with thigh-deep wet guano having the look and feel of pea soup. We waded in, pairs of us following different mine shafts. Suddenly a large bat flew past. It looked like a vampire, and we hurried back through the mine toward the others to let them know we had found a bat.

We unfurled a bat net to try to block the mineshaft. These nets, which feel much like hairnets, are so fine that the bats' sonar does not detect them and the bats fly into them. Usually nets are hung over water or across trails, but there was no place to hang a net in the rocky mine. I decided to station myself on the bat side of the net while my companions held the net up across the tunnel to trap the bat as it flew toward them. Three others headed deeper into the mine to drive the bat toward the net. My job was to grab any bat that flew into the net. Soon a large bat flew swiftly toward me and became entangled in the net. I grabbed it—my first vampire—and held it until it could be put in a bag.

A second vampire bit the student who was holding the bag, the bat's teeth so razor sharp that he did not know he had been bitten until we noticed blood trickling down his hand. It took a long time to control the bleeding because the vampire bat's saliva contains an anticoagulant that impedes blood clotting. The student underwent the painful series of rabies shots as soon as we returned to the United States because the bat escaped after biting him and thus could not be tested for rabies.

We collected many mammals that were new to me on the trip, worked in a wide diversity of habitats, and added to the museum's research collections. I found fieldwork enormously enjoyable.

We returned to Albuquerque exhausted but exhilarated by our experiences in Mexico. Classes started within days of our return. A week later I developed a cough and fever. I assumed that it was a cold and paid little attention to it. But the fever and sore throat got worse. Every day I would drag my increasingly ill body to the university, where I would cough, wheeze, shiver, and sweat. No one else who went on the trip was sick.

My symptoms worsened and eventually could no longer be ignored. Within a week my temperature climbed to 102°, then 104°, at which time I went to the doctor. He listened carefully to my history and concluded that I had been infected by something in Mexico. After many blood tests and X-rays, he decided that I probably had contracted leptospirosis (swamp fever, a bacterial infection) in the vampire mine, its thick, wet guano being an ideal medium for harboring pathogens. He began strong antibiotic therapy. After a week my temperature had climbed to 105° and I was losing my voice because my throat was so inflamed.

I returned for further tests. One of the doctor's partners specialized in infectious diseases and was brought in to consult. I underwent further blood tests and additional chest X-rays. I had been X-rayed initially, but this time the X-rays were different. My lungs, which had been a nice, clear gray color on the first films, were now streaked with white. You did not have to be a radiologist to see that great changes had taken place in a week. I asked if I might have contracted malaria.

"If this is what we think it is, we might all wish it were malaria." This remark was not comforting.

"What do you think it is?" I asked.

"We think you have histoplasmosis, but a particularly virulent form." My disease now had a name, but not one that I recognized.

"If we are correct about what this is, you will have to go to the hospital."

"Well, I have my first-quarter exams coming up soon. Maybe I could go in the week after next. I can't skip the exams."

"I don't think you understand," he said, "I was thinking of having an ambulance transport you to the hospital. You need to be admitted immediately."

"What about school?" I asked.

"We don't know how this disease will respond to treatment, but school is over for you for the time being."

It is hard for a twenty-one-year-old to comprehend serious illness. Clearly I was sick, as my blazing fever, ulcerated throat, and rapidly diminishing ability to speak attested. But could I be "hospital sick?"

The doctors had turned their attention from the abandoned silver mine awash in wet guano as a likely source of the pathogen to Cueva del Tigre, with its tons of dried guano that I had waded through. Leptospirosis, being a bacterial infection, would have responded to the antibiotics. Histoplasmosis is a fungal infection and the spores of the fungus are associated with dried bird or bat droppings. Pathogenic fungi are notoriously difficult to kill and do not respond to antibiotics. In most cases histoplasmosis manifests itself with mild, flulike symptoms, and this form of the infection may be confused with the flu. Flulike histoplasmosis is seldom fatal. After running its course it disappears—in most cases. The much rarer form of progressive disseminated histoplasmosis is another matter.

There was not a lot of information available on the disseminated form of the disease. As I examine my 1992 edition of the *Merck Manual* (a physicians' disease reference book), I find that there is more information available than there was in 1966, and my case may have led to some of the comments in the later edition. The disease spreads from the lungs and moves into the body like a cancer. It affects the liver, spleen, lymph system, adrenal glands, and, "less frequently" (this statement may refer to my case) "causes ulcers in the throat." The treatment for the disease in 1966, which was not very successful, involved injecting heavy metals into the body, poisoning the patient until perhaps the fungi were also poisoned. Today, with a better treatment available, the disease still has a fatality rate of 90 percent. In 1966 it was almost invariably fatal.

I lay in the hospital for over a week, becoming sicker each day. The ulcers in my throat were so bad that I could hardly speak at all. Alcohol rubs and various medicines were unable to keep the raging fever in check. I had to withdraw from all classes except one, ornithology, which I loved as much as I loved mammalogy. Within a few days of dropping the classes I received my draft notice.

This was during the early days of the Vietnam War and I was not eager to be sent off to fight a war that was already unpopular at home. I became depressed about the whole situation. My quiet, uneventful, and enjoyable life as a student in biology had come to an end. I now faced a world full of hardship and death. I would be drafted before I could finish my studies, assuming that I lived.

I was lying in bed, hot and uncomfortable with my fever raging, holding my draft notice, when the doctor came by to check on me. I pointed out that I was in a very difficult position and asked if he could call the draft board and let them know that I might need an extension, since I did not know whether or not my treatment would be successful.

"I'll take care of it," he said. "They won't bother you until we get this disease cleared up."

One Friday afternoon my physician said that if I did not improve over the weekend he would begin a different treatment on Monday. He had previously let slip that he would only embark on this thirty-day treatment if there was no other hope. The news that we would proceed to this phase of treatment, which was described not only as painful to undergo but as unlikely to be successful, depressed me. Now I had dropped out of school, gotten drafted, and could die, all because I had been looking for a bat in a cave in Mexico. Who would have believed such a scenario?

There is a cartoon that is popular in scientific circles that shows a mathematician solving an exceedingly complex problem that fills a blackboard with convoluted equations. Just before he derives a simple, elegant solution, he writes, "And then a miracle happens." Such was the case with my wildly disseminated histoplasmosis. The disease had led to seemingly hourly blood tests, numerous X-rays, and agonizing bone marrow examinations done with sternal punches—the physician climbs onto the bed, straddling you, and pounds a huge needle into the sternum that sucks out the marrow.

It was on a Sunday night when my miracle happened and I was awake for

A cave inhabited by Brazilian free-tailed bats *(Tadarida brasiliensis)*. The investigators wear dust masks to avoid contracting histoplasmosis. (Photo by M. A. Mares.)

it. I could not sleep knowing that the dawn would bring a dangerous treatment that would probably not cure me. As the hours crawled by and I reviewed my life up to that point, I knew that there was still much I hoped to do. Yet it seemed that I would never be able to go into the field again, which I yearned to do. I wanted to be a field biologist. Every time on the trip to Mexico that we found animals that were new to me—or animals I knew about but I was encountering in new habitats—I felt as if a curtain were being parted. I had entered biology with such great ignorance about animals that each day in the field was truly enlightening. I could not think of a more enjoyable way to go through life than learning new things about nature each day and living in the field with friends. But my life was slipping away, and even if I survived the histoplasmosis, I would still have to survive Vietnam.

As these thoughts raced through my head, I suddenly felt my body suffused with what I can only describe as a glow. It was the most extraordinary sensation, as if a beam of light had filled each cell. I felt my fever break and my throat begin to heal almost instantly. I knew without a doubt that my disease was receding and that there would be no need for treatment on Monday. I had somehow, inexplicably, overcome the disease.

As a biologist I know now that what I was feeling was the sudden action of my immune system overcoming the fungi that had spread through my body—the defeat of the pathogen attacker by the immune system defenders. I stayed awake all night waiting for the doctor so that I could tell him what had happened. He arrived early the next morning. I excitedly told him that the disease had retreated overnight and that there was no need to proceed with the new treatment. My exhilaration and the fact that I could talk so easily impressed him, but "skeptical" would not begin to describe the look on his face. He felt my forehead and noted that the fever had broken. He looked at my throat and saw that the ulcers were healing. He said that he had never seen anything like this and ordered another chest X-ray. The lungs, which only a day earlier had been coated with the telltale white covering of the fungus, were almost clear.

The physicians were stunned. I had experienced a spontaneous remission, which, had it occurred just as a faith healer had touched my head or a shaman chanted incantations around a fire, would be termed a miracle. Suddenly I was cured. I left the hospital several days later. Specialists came from Miami and San Francisco to run tests on the "Spontaneous Remission" (me). My lungs were washed, and I was poked, prodded, and X-rayed many times before they were satisfied that I was healthy.

After returning home I received a letter from the Selective Service Office. Because of my serious lung condition, I had been given a permanent 1-Y deferment. The Office was unsure whether the fungus had damaged my lungs and other organs. I was immune from the draft. I could proceed with my education and with my fieldwork. Now *that* was a miracle.

The Winding Path to Field Biology

If I had not been admitted to these studies it would not have
been worth while to have been born.

Seneca, *Naturales Quaestiones*, first century

Following my recovery from histoplasmosis, I completed my undergraduate studies at the University of New Mexico and followed these with a master's degree from Fort Hays State College in Hays, Kansas. At Hays I was trained in ecology, and my thesis dealt with the population biology of cotton rats on the prairies of western Kansas. At Hays I also received a strong grounding in all of the "ologies," teaching labs in mammalogy and herpetology, and taking courses in advanced ornithology, ichthyology, plant and animal ecology, wildlife ecology, field biology, human ecology and conservation, the biology of the Southwest, and comparative physiology. After applying to several universities for admission to the Ph.D. program, I was accepted by the University of Texas at Austin. The only reason I applied to that school was that there was no admissions fee required with the application. I entered the graduate program of the Department of Zoology in 1969, arriving the day Neil Armstrong walked on the moon.

I took a class in biogeography that was taught by Dr. W. Frank Blair. Blair was a legendary figure who had written several important books, published hundreds of scientific papers, helped develop the field of communication ecology in amphibians, and done seminal work on the population ecology of small mammals and lizards. He was also helping coordinate the International Biological Program (IBP), a multifaceted approach to research, involving hundreds of scientists throughout the world, with the ambitious goal of achieving a basic ecological understanding of all the world's ecosystems.

Blair was a shy man who was uncomfortable in front of the class. One day as he began class he said, "If anyone would like to go to South America to do their doctoral research, I have money to support you." There were perhaps twenty graduate students in class, and after he made the statement he continued with the lecture as if nothing unusual had happened. For me, however, the opportunity was tantalizing. Could field research in South American be in store for me? I was a brand new graduate student who had not yet identified a major professor and who had a master's degree from a small Kansas college. Surely I had no chance to be chosen. I could not concentrate on the class for I was certain that as soon as it was over the entire group would beg to be sent to South America. They would all, I thought, rank ahead of me.

The class ended and I held my breath to see how many students would rush forward. No one did. Should I say something? I had hardly spoken to Dr. Blair during previous classes. If Blair was shy, I was equally so. I asked him if I could speak to him about the South American project. "Sure," he said, "come on down to my office." We did not speak as we walked to his office. I decided to be as direct as possible.

When he sat down at his desk, I said, "Dr. Blair, I am ready to go to South America to do research, and I am willing to work on reptiles, amphibians, mammals, birds, or fish, although I don't particularly like fish."

He replied that the research was on mammals.

"I love mammals and I did my master's work on rodents."

He said the work would be in the Monte Desert of Argentina.

"I love the desert and I am ready to go to Argentina."

"Okay."

I asked when I would go and he said it would be in the fall of 1970. I had made one of the most important decisions of my life in just one hour, without having a chance to discuss it with my wife, Lynn. I had no research plan. No scientific questions had been articulated that required this overseas research. This is not the way a graduate student is supposed to do a research project and later, as my own career developed, I never would have allowed this to happen to one of my own students. No matter how you look at it, I was incredibly naive—but I was going to Argentina.

Blair wanted me to spend all of my time preparing for my graduate exams, which included an oral exam covering all areas of biology. But I had applied for admission to the Organization for Tropical Studies (OTS), in order to take a class in tropical ecology limited to twenty students chosen from universities throughout the country. It was a grueling eight-week course taught in the rainforests of Costa Rica. Students lived in the field for the two-month period. A friend had taken the course and found it a life-changing experience. Blair said that if I could get into the class, I could go. I was accepted and left for Costa Rica in February 1970.

OTS was all that I had hoped it would be and much more. The students were from Stanford, Berkeley, Cornell, Harvard, Michigan, and other prestigious institutions. Well-known ecologists, tropical biologists, zoologists, and botanists taught the classes. Class days were long, beginning before dawn and ending after dark. We studied everything from plant ecology to insect behavior to interactions between plants and animals. In addition, we conducted our own research projects on organisms of interest to us. My projects involved the ecology of tadpoles and bats, and I later published two papers derived from these projects.

Studying bats meant that after the class day was over my research would begin. Another student and I collected bats at night, releasing many, but retaining some for further study. After identifying them, we recorded their reproductive status, the shape of their teeth, and other morphological characteristics. We were comparing bat communities among the major habitats of Costa Rica. Going without sleep became routine. Those of us who worked at night became a special subgroup of students, commando-type biologists who did not need sleep. I doubt that we averaged more than four hours of sleep a night during the course. I was twenty-five years old and loved the regimen. Everywhere I turned there were new things to learn and new ways to look at nature. During the course, we worked and studied in all the major habitats of Costa Rica, from cool and damp oak cloud forests above 12,000 feet to hot lowland rainforests at sea level. We even spent a week in the Caribbean working on coral reefs.

One night a classmate and I were studying bats at a site ten miles from the base camp. There was a shortage of vehicles, so we were delivered to our research site in the tropical rainforest in the late afternoon and would be picked up in the morning. We placed a series of bat nets in the remote forest and waited for the bats to fly. It was a gorgeous tropical night, the air so moist and smooth that it seemed to be caressing us. We could hear the soft fluttering wingbeats of the bats as they flew. But then suddenly the wind began to blow hard in the treetops a hundred feet above our heads. It was odd to hear the wind blowing in the thick canopy but not feel it at our level. The denseness of the canopy kept the wind from penetrating to the ground.

In our lectures we had learned that trees in the tropics allocate enormous amounts of energy to the aboveground plant parts—trunk, stems, and leaves—in the effort to outgrow and outcompete other trees for light in the upper reaches of the dark forest. Tropical trees allocate less biomass to root tissue than do their temperate counterparts, as Marius Jacobs later noted in his classic work on the tropical rain forest. As a result, winds that would not normally cause temperate trees such as conifers or hardwoods to topple will knock down these less firmly rooted tropical giants. We were not thinking along these lines as we sat on the tranquil forest floor waiting for bats to enter the nets and listening to the wind high above our heads. The wind increased in intensity and soon we began to hear the giant trees fall. As we listened, the sounds came closer.

There was little we could do about the falling trees, so we continued to check our bat nets. One net was set near a barbed wire fence. Others were placed in openings in the forest. I had set yet another across a creek where I hoped to catch fishing bats (Noctilio leporinus), large fish-eating bats that pluck the fish from the water with their fishhook-like claws, using sonar to detect the ripples in the water caused by the fins of the fish. Many bats were flying into the nets. As the wind continued to increase and as the trees continued to fall, we began to worry. Suddenly there was a loud crack nearby.

We were surrounded by hundreds of huge trees that were invisible in the dark. Somewhere nearby one of these monsters was about to topple.

Tree ferns in the lowland tropical rainforest of Costa Rica. The 20-foot-tall tree ferns are much smaller than the 100-foot-tall trees in the background. (Photo by M. A. Mares.)

My classmate yelled, "Let's get out of here," and began running along the fence line like a lizard scurrying from danger.

"Wait," I shouted as I chased after him, "we don't know which tree it is. Let's wait for another crack."

I caught up with him and we stood panting next to a massive tree well over a hundred feet tall. Suddenly we noticed that the giant tree seemed to be undulating.

"Run!" we both shouted, and tore back down the trail the way we had come. Behind us, we heard the tree begin to break and fall. We raced along the narrow trail as the great tree fell and felt the wind from its leaves as the crown slammed to the ground only a few feet behind us. We had escaped by inches. As we stood there gasping and clapping each other on the back to celebrate our narrow escape, we heard other cracks echoing like musket fire through the forest.

It was a long night. As I checked the net placed across the creek I heard the loud snap of a tree beginning to fail. I cast about frantically trying to locate the tree. On the bank nearby was a tree with a trunk perhaps 5 feet in diameter and with giant branches extending out over the net. It was shaking violently and was about to fall on the spot where I stood removing bats from the net. I was waist deep in water. I began to slog toward the bank, but it was difficult to move because my feet were mired in mud.

A slow, nightmarish ballet ensued. I would pull my feet out of the mud with a sucking sound, only to plunge them back into the ooze with each step. I could only move in slow motion when I was fleeing for my life. When I finally reached the steep muddy embankment, I began to claw my way up the side, only to slide back into the water. The tree continued to crack and shake. Scrabbling frantically at the muddy bank, I at last pulled myself onto dry land and ran into the forest just as the huge tree crashed down on the net.

About a dozen trees fell very close to us that night. We did not sleep at all because of the falling trees. But we did catch many bats and survive the tropical equivalent of a mortar barrage. For a week after the night the trees fell, though, we jumped whenever anyone stepped on a branch and made it snap.

When the OTS class was over, I returned to Texas and prepared for both my research project and my graduate exam. I had planned a project that would investigate the ecology of competition among the desert rodents of the Monte. I would determine how species managed to coexist in the arid desert and compare my results to those that had been obtained over many decades of research in the Sonoran Desert of Arizona. My project was based on competition theory, the mathematics of which were then just being developed by Robert MacArthur and his students. Some recent research on desert rodents in North America had developed several hypotheses about rodent coexistence, noting that deserts seem to support extraordinarily high numbers of coexisting species that forage on the same resource: seeds. I hoped to determine whether the small mammals of the Monte Desert exhibited similar patterns of diet, coexistence, body size diversity, habitat selection, and other ecological factors that seemed to be important in reducing competition and permitting coexistence. I had never been to the Monte Desert and no ecological research on mammals had been conducted there, but I read everything I could find about the Monte and about the mammals of Argentina. I was convinced that I could apply theories developed in

North American deserts to that desert. After all, how different could the two deserts be?

The study of ecology in the 1970s was a search for patterns in nature, for if nature is not predictable it would be impossible to develop a grand theory that explained why and how nature functioned as it did. If every habitat was unique in the way it functioned, scientists would have to study each habitat individually in order to understand it, manage it, exploit it, or protect it. There was an influential school of thought that maintained nature could be reduced to a few simple mathematical formulas or computer models.

I was about to study the mammals of two deserts whose vegetation looked similar and that had similar landforms and climates. Was there a profound pattern to nature? Would the mammals of these two deserts that were more than 6,000 miles apart—mammals with different evolutionary histories— look, act, and function alike? Could one understand all deserts by under- standing only one desert? I was convinced that mammals in the Monte would be both abundant and rich in species, just like those in the Sonoran Desert. I was sure that similar adaptations would characterize these unre- lated species, even though they were only distantly related. I felt certain that I would find that evolution worked in a predictable manner. Nature was about to teach me a lesson in humility. In my cloistered and sublime igno- rance, almost nothing that I hypothesized about mammal ecology in the Monte Desert was correct.

Crossing the Chacoan thorn scrub and then the Monte Desert in the heart of the Argentine summer in a Citroen 3CV was memorable. The car was a small, uninsulated tin box with a cloth roof. The windows opened by folding upward. My wife and I traveled from the tall scrublands and rich farmlands of Córdoba, a large city in central Argentina where we had purchased the car, across great salt flats surrounded by tall candelabra-like cacti and salt- loving vegetation. Mostly we drove on dirt roads and periodically had to work our way across arroyos that had recently been flooded.

As we came through the cactus- and bromeliad-encrusted steep hills of the rocky desert canyon that connects the Chaco, the Great Thorn Forest, with the Monte, we entered the upper slopes of the desert. The similarities with the Arizona desert were obvious. The tall cardón cacti *(Trichocereus)* looked like the saguaro cacti of Arizona. They even had holes in them that must have been made by woodpeckers behaving like the Gila woodpeckers that drill holes in the saguaros. Creosote bush covered the valley flats, as it does in New Mexico and Arizona. There were acacia *(Acacia)* and mesquite *(Prosopis)* trees, and even paloverde *(Cercidium)*, just as in the hot northern deserts. It seemed that we had arrived back in North America after traveling more than 6,000 miles to the south.

Monte Desert vegetation with cardón cacti *(Trichocereus terscheckii)*. (Photo by M. A. Mares.)

Finally, after negotiating a tortuous mountain road, we entered the southern end of the Bolsón de Pipanaco, an isolated valley in Catamarca Province, in northwestern Argentina. As we dropped into the valley en route to Andalgalá, the site chosen as the base for the IBP project, we encountered a haboob, an unusual windstorm that occurs in deserts when thunderstorms are near. In a haboob dirt is lifted aloft to a height of several thousand feet and the air becomes opaque, almost colloidal. I could no longer see clearly because dirt and dust piled up on the windshield. We inched along, able to see only a short distance ahead. I stopped the car periodically to allow the air to clear enough for me to see the dirt strip that served as a road. Outside the car, we could breathe only dust.

After driving through the vast dust column for some time, we suddenly emerged into clear air and saw above us crystal blue skies with puffy white clouds. The desert plants shown with the brilliance of the summer sun and the cacti were highlighted against the pellucid sky. I stopped the car and looked back at the sharply defined vertical wall of dust extending upward for thousands of feet, seeming to stand motionless in the desert. The top of the dust cloud was overlain by blue skies. Nature had prepared a rare welcome for us—one that I have never experienced again in decades of desert research.

When we arrived in Andalgalá, the town was a sleepy desert oasis. The lack of television and radio was welcome, for the old ways were seductive. People rose early and businesses opened at eight o'clock. At noon everything but the cafés closed for a long siesta that lasted until six o'clock in the summer, but was shorter in the winter, when dark came early. Lunch was the big meal of the day, and after lunch, especially if much wine had been consumed—and usually it had—people passed the heat of the day napping, visiting in the sheltered patios of their homes, or sitting in the shade of their small grape arbors. There was no air conditioning, but the thick adobe walls of the houses kept the indoor temperature cool even during the hottest weather.

After the siesta, stores opened sporadically and did not close again until nine o'clock. Cafés did not start serving dinner until ten o'clock. After dinner, and sometimes before, many of the townspeople walked to the central plaza, where they promenaded around in a circle, greeting their fellow citizens as they passed each other. The young women and their chaperones (mothers or aunts) walked in a counterclockwise direction, and the young men walked in a clockwise direction. Each time they passed each other, they politely said *adios*—literally meaning "go with God"—but a word used to say both "hello" and "good-bye." Hundreds of people did this every night. Entire families strolled along together greeting other families. Every young woman primped for the promenade because that is where she probably would encounter her future spouse. Those casual greetings were the first step in measuring each other as potential mates.

Life in Andalgalá was neither bucolic nor even especially serene. It was traditional, however, and its people engaged in many practices and followed customs that had been established in Spain or Italy long before Argentina was colonized. The town was isolated—when the national government was overturned in a coup d'état the news did not reach Andalgalá for three days—and it had persisted in seclusion for centuries, its people bound together by traditional social rituals. Things were about to change.

Not only had the International Biological Program chosen Andalgalá as its base of operations, a decision bound to bring many changes, but a copper mine was being dug nearby, promising to bring jobs and wealth to the town. Andalgalá had never had any major economic activity during the several centuries of its existence, so the new mine offered the prospect of fundamental changes in the town's way of life. Both the research project and the mine brought a great diversity of people to Andalgalá from all over the world—biologists, geologists, missionaries, miners.

Each person who arrived from another country had a unique story of how he or she happened to be in that remote village in the Monte Desert. One of them was a man from Scandinavia I befriended who was a foreman at the copper mine, situated high in the mountains above the town. Lars, as I shall call him, was doing difficult, dangerous, and demanding work. Even getting to the mine was a challenge. Miners had to ride from the town to the mine on mules—an eight-hour journey over trails only a foot wide, with sheer drops of thousands of feet on the outer edge. Occasionally, a mule heavily laden with equipment fell to its death, and landed so far down the mountain that no sound was heard when it hit the bottom.

The arduous mule trip to the mine, at an elevation of 15,000 feet, was followed by the grueling work of mining for fourteen straight days. This was followed by a week off from work, and mule trains brought the miners to town for relaxation. Some of the miners, and certainly Lars, spent all of their time away from the mine drunk, preferably in a whorehouse. Alcohol was not allowed at the mine, so miners were on enforced good behavior for two weeks at a stretch.

Lars was a big, rugged man who not only loved to drink, but also loved to fight. Once while we were sitting in a bar in Andalgalá, he asked me very courteously if he could punch me in the face so that I would sail into the table behind me where a group of local toughs were drinking.

"Why would you want to do that?" I asked.

"That will get them to start a fight with you and then I can join in," he said with ill-disguised anticipation.

I imagined his big paw hitting me in the face and convinced him that I was not in a mood to fight, much less to have him knock me into the next table. He was as big as a refrigerator and tough as nails. He would have killed me. I suggested that we finish our drinks and leave.

"I would like to have one more good fight before I go up to the mine," he said hopefully.

"Good luck to you then, Lars," I said, "I need to get some sleep. I hope your fight goes well." I left feeling sorry for the local boys.

Lars did not get into a fight that evening, but he drank heavily the entire night and in the morning was so drunk that he could hardly walk. He showed up at the place where the mule train departed, a 5-liter demijohn of red wine clutched in each of his bearlike hands. If nothing else, miners are practical. Lars's coworkers tied him to the mule so he would not fall off and then tied a wine jug to each hand so that he could drink from them on the journey. The mule was smacked on the haunches and began the eight-hour climb into the clouds. If the mule did not fall into the abyss, Lars would have all night to recover before beginning two weeks of hard work and sobriety.

Lars had traveled the world, as many miners do. He had a black family somewhere in Africa that he missed terribly and he probably had other families on other continents. Occasionally he cried about them, usually after he had been drinking heavily. One morning I ran into him as he was returning from Tucumán, where he had spent the previous week in a whorehouse. He looked especially forlorn.

"Good morning, Lars, how are you doing this morning?" I asked, and then added, "Actually, you're not looking well. Is something wrong?"

"Ah Mike, I caught gonorrhea again," he said dejectedly. "I guess I will have to take care of it when I come back down the mountain." It was the second or third time that he had caught one venereal disease or another since arriving in Andalgalá.

During the two years my wife and I lived in Andalgalá (and another two years when we visited regularly), we saw it survive intense excitement at the prospect of a great copper mine, a time when foreigners descended on the little community like locusts on a crop. The mine never panned out, not yielding enough copper to justify continuing its existence, although for a couple of years it was a source of unprecedented excitement and drama in the little desert town. The town with no phone or radio or television was suddenly inundated with Americans, Norwegians, Swedes, and Chileans from whom the money flowed in an apparently endless stream. Helicopters were constantly roaring off into the high mountains, taking company officials to visit the mine. Short-wave radios were set up in town so the miners below could talk to those high on the mountain.

Visions of sudden riches danced in the townspeople's heads. Hopeful entrepreneurs who had never before had more than a few pesos at one time hatched get-rich-quick schemes by the dozen. Grandiose plans were tossed about. New roads were planned and the possibility of building new restaurants and hotels was discussed. The people of this impoverished region dared to dream of becoming wealthy. Gradually, however, hopes faded as no rich veins of ore were discovered. Eventually the noise from the helicopter rotors was carried away on the desert wind and the miners departed. No new businesses were built. No fortunes were made. The poor town in the isolated valley settled back into its somnolent existence and the great mine was only a memory.

By 1974 things had changed radically. Radio and television had scaled the high pre-Andean chains and arrived in the small town. The evening promenade had largely disappeared; people sat glued to their TV screens, even though few programs were offered. Television sets were left on in cafés and bars whether or not anyone was watching or listening and whether or not there was a picture. News from Buenos Aires, the sprawling capital of eight million people lying more than a thousand miles to the east, now inundated the town. For the most part the news had no relevance to the people of Andalgalá. Of what consequence could it be that a wreck had occurred on a freeway along the coast of the Atlantic Ocean? There were no freeways in Andalgalá or even paved roads outside the town limits. What did a flood in Buenos Aires matter to people who had not experienced a flood in a century of desert living? Crime, traffic problems, and fires were all experienced vicariously and in color. Some of the cafés closed down when people stayed home to watch television.

The small theater that mostly played American movies was closed down for the same reason. When it was operating, the sound system produced audio tracks so garbled that no one could have understood them, but that mattered little since the townspeople did not speak English. What the public would not tolerate was a break in the film, and the film did break three or four times during each showing. As soon as it happened, the lights would come up and hordes of vendors who sold an array of snacks and who were eagerly awaiting the inevitable break in the film would appear in the aisles. The audience would begin whistling, stamping their feet, and clapping loudly until the harried projectionist was able to tape the pieces of the film together again and proceed with the movie.

The communications age had arrived in Andalgalá very quickly, and the delicate threads of social norms that had been spun on another continent and had been nurtured for centuries in Argentina were irreparably broken virtually overnight. Andalgalá was changed forever.

Argentina in 1970 was in many ways similar to the New Mexico of my childhood. I was quickly seduced by the country and still consider it a second home, but it is a difficult country to love. Its history is filled with unstable democratic governments and cruel dictatorships. In the years that followed my arrival, as I conducted research throughout the 1970s and 1980s, and on into the 1990s, I would see both the best and the worst sides of Argentina. The nation grappled with military juntas, Nazi-like governments, economic instability, and eventually underwent a slow return to democracy. At times during this period its economy was among the worst in the world.

During my time there the nation suffered through an internal "Dirty War." Guerillas tried to overthrow the government, and both military and police forces were unleashed on the country's civilians. During this chaotic period the government killed more than 10,000 of its citizens. They were dropped into the ocean from helicopters, exploded on dynamite kegs, and electrocuted. Pregnant women were kept alive until they gave birth, then their children were given to childless military families and the mothers were killed. Areas where we wished to work would be declared free-fire zones and anyone found there could be shot. Trying to do field research in these conditions was difficult.

Beyond the dangers from guerillas and the police, the economy was plunged into chaos. In late 1970, 400 pesos equaled a dollar. Within a few months, a dollar was worth 1,200 pesos, meaning that if you had bought a car for $2,400, as I had, it was worth less than $800 two months later. The depths of the country's economic problems could be tracked by following the currency devaluation. Soon after hyperinflation began, the money supply had to be reprinted, for there was not enough room for all the zeroes that needed to be printed on the bills. At one point, there were both new and old pesos in circulation, and you had to keep track of which was which: 1,200 old pesos equaled 1 new peso.

It cost more to mint peso coins than the coins were worth, so coins disappeared. Then paper pesos began to be hand-stamped to show their new value because they lost value too rapidly to be reprinted. Armies of clerks stamped each bill then returned it to circulation. Soon four zeroes were added to the bills, then another three. Finally the zeroes dropped away and the austral was born.

The austral was worth more than a million of the original old pesos, and for a moment each austral was worth a dollar. But the zeroes, even on the austral, continued to multiply. Its linkage to the dollar was soon only a memory. Eventually, with the demise of the austral and the rebirth of the peso, 10 million old pesos were worth one new peso, then 100 million. At one point, if a single 1980s dollar had been converted to the old pesos that were in use when I arrived in 1970, I could have purchased thousands of vehicles. Put another way, it took a fleet of 1970 cars to buy a cup of coffee in the 1980s.

It takes little imagination to understand what this rate of inflation—the highest in the world—did to the lives of the people. Currency was worthless. Salaries were paid on the first of the month, and since the money lost value by the minute, people rushed to grocery or department stores to convert all of their money into food or hard goods. The prices of goods were not even listed in the stores, for they would change by the time a shopper got to the checkout line. Shopping became a lottery.

At the beginning or end of a field season, I never knew if a hotel room would cost one dollar or a hundred. Would a meal cost 50 cents or $20? Would a car cost $800 or $40,000? Financial problems always played an important role in my research. But problems with money were not new to me. During my first few years in Argentina, my wife and I often did not have enough money to conduct field research, lacking funds for gasoline, car repairs, food, or equipment. Once for more than a week we lived on the oatmeal that we used as bait since we had no money for food. Each day we would put some oatmeal aside for the traps and another portion aside for ourselves. Such uncertainty of funding was a part of all field expedition planning until the 1990s, when the currency stabilized and we had access to more funds.

But we managed to continue our field research even during the darkest periods. Our two sons were born in Argentina, on different field trips. Spain's "Land of Silver" had become a permanent part of our lives. But all of that lay before us as we settled into the town that would be our new home. In January 1971 I began my doctoral research on the ecology and evolution of the mammals of the Monte Desert.

The Desert at the Bottom of the World

Desert regions, despite the harshness of their environment, can produce some of the most picturesque scenery in the world. Sometimes the landscape is rugged; often it is attractive in the aesthetic or artistic sense, in the textures created by patterns of light and shadow amongst the sand dunes. There can be a serene beauty in the light of dawn, which, though lost to stark realism in the midday sun, returns, changing by the minute, through a swiftly descending dusk to the unequalled clarity of a starlit desert night. No two deserts are the same, and within any desert lie an infinite number of scenarios. Some are on the grand scale, others much more of a cameo, but all reveal some element of the very special nature of deserts.

Jim Flegg, *Deserts: A Miracle of Life*, 1993

South America, unlike North America, which was once connected to Eurasia, was a huge island for much of the time when mammals were evolving on that continent. When the earth's continents began to move apart in the Cretaceous period about 100 million years ago, South America split off from Africa, with the Atlantic Ocean gradually expanding and increasing the distance between the two land masses. Mammals were present on both continents when the breakup occurred, and some faunal exchange may have taken place while the continents were still close to each other. After 15 million years had passed, 300 miles of ocean separated

Africa and South America. Faunal exchange between these southern continents became uncommon.

One group that appeared early in South America is the caviomorph rodents, guinea pigs and their relatives. Caviomorphs came from Africa, but when they reached South America they underwent a great diversification. They evolved to fill many niches over a period of 45 million years on the huge island of South America, ranging from large hippo-like species to small gopher-like animals. The largest rodent in the world belongs to this group, the capybara (*Hydrochaeris hydrochaeris*), an animal weighing more than 100 pounds that lives in rivers and marshes and behaves like a small hippopotamus. The tucu-tucos are gopher-like burrowing rodents. The nutria (*Myocastor coypus*) is a large swamp rat. Cavies (members of the guinea pig family) are similar to ground squirrels, and other groups, such as the octodontids, are like woodrats. Plains vizcachas (*Lagostomus maximus*) are 15-pound colonial rodents that behave like enormous prairie dogs. The Patagonian "hare," or mara (*Dolichotis australis*), looks like a cross between a deer and a jackrabbit. Such diversity indicates a long period of time during which the animals adapted to the habitats as the habitats themselves formed.

The other major group of rodents in South America is the sigmodontines, which appear to be related to North American rodents such as the woodrat, the deer mouse, and the grasshopper mouse. If the North and South American "sigmodontines" (named after a rather primitive species, *Sigmodon hispidus*, the cotton rat, which occurs in both North and South America) are in fact related, the relationship would be very distant, extending back millions of years to a common ancestor. The colonization of South America by sigmodontines is assumed to have taken place when the two continents became connected by the Central American land bridge about 2 million years ago (at the same time the porcupine, a caviomorph rodent, moved northward into North America, where it still occurs today). Some scientists believe the sigmodontines colonized South America millions of years earlier, but the fossil record supports an early arrival in South America for the caviomorphs and a much later arrival for the sigmodontines. Since the South American deserts are as old as the North American ones, the guinea pigs and their relatives had a much longer exposure to arid habitats than did the invading field mice and their relatives.

One of the things that the International Biological Program was examining was whether or not different floras and faunas evolved in similar ways when exposed to similar environmental conditions. To determine how similar the mammals were between the North American Sonoran desert and the South American Monte, I would have to measure such things as numbers of species, population density, habitat selection, major niche types, general food preferences, degree of coexistence, degree of physiological and anatomical similarity, and other parameters of desert life. Most of these data were available for North American desert mammals, but nothing had been published on Monte mammals. Answering the question for mammals was complicated by the fact that the two very different lineages of rodents, the caviomorphs and the sigmodontines, had entered South America during different periods, the Eocene of 45 million years ago for caviomorphs and the Pliocene-Pleistocene boundary of 2 million years ago for sigmodontines. I had to find a way to develop comparative techniques that would assess how similar the mammals were between two disjunct deserts. I had to learn as much as I could about each of the desert species in Argentina so that I could compare them with their possible North American counterparts or determine that they had no counterparts in that distant desert.

In Old World deserts such as the Sahara, another group of rodents that was only distantly related to kangaroo rats, the jerboas (for example, *Jaculus*), seemed to be exact ecological analogues of the kangaroo rats, being bipedal, having a long, tufted tail, foraging on seeds, and so forth. North American deserts had bipedal rodents and no major desert seemed to lack them, so with this great bias in mind, I hoped to find bipedal rodents in the Monte Desert. Why should evolution work any differently in South America than it appeared to elsewhere?

South America was not an unknown continent as far as its mammals were concerned; field biologists had labored there sporadically for more than two centuries. Charles Darwin had arrived in Argentina on the *Beagle* in 1831 and had made some of the earliest mammal collections. In no published works had I ever found reference to a bipedal species, but such a discovery would not be out of the question. If convergent evolution were a powerful force in rodent evolution, why would there not be an ecological equivalent to the kangaroo rats in the Monte?

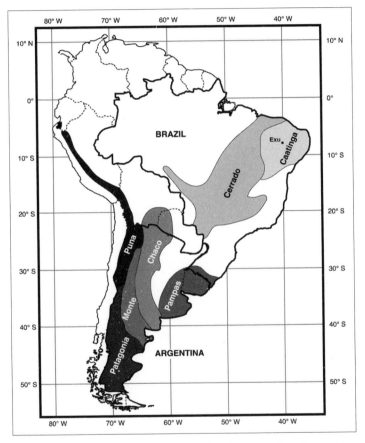

The arid and semiarid areas Argentina and Brazil. (Map by Cartesia MapArt, modified by Patrick Fisher.)

The Monte Desert of western Argentina is at first glance the mirror image of the Sonoran Desert far to the north. Two photographs of the deserts laid side by side might seem to have been taken in neighboring valleys of the Sonoran Desert, rather than on different continents in different hemispheres. The

Monte extends more than 1,200 miles from just south of the Bolivian border through much of western Argentina in the lowlands east of the main Andean mountain chain and west of the pre-Andean chains. In the northern half it is a very narrow desert, seldom exceeding 100 miles in width and always nestled between massive mountains. These great mountains, with peaks exceeding 20,000 feet in height, keep the moisture-laden clouds from eastern Argentina or from the Pacific Ocean from reaching the isolated desert valleys. From Mendoza Province southward, the Monte is limited on the east by thorn-scrub habitats. To the west lie the Andes. Over most of the desert, elevation ranges from 3,000 to 4,000 feet above sea level, but at its southern tip on the Peninsula Valdés the Monte abuts arid Patagonia and reaches the sea.

Northern Monte Desert habitat in La Rioja Province, Argentina, with the Sierra de Famatina in the background. (Photo by M. A. Mares.)

My research would come to show that although the Monte and Sonoran deserts look alike, from the viewpoint of their mammals they are quite different. In the northern Monte, the desert supports a fauna that, before their numbers declined owing to hunting pressure, used to include the guanaco (*Lama guanicoe*), a 200-pound camel that still occurs from the highest Andes to the shores of the Atlantic Ocean in southernmost Argentina. Guanacos are rare on the upper slopes of the Monte, but they are common in the southern Monte and in Patagonia. Similar camels had once roamed the American Southwest but had disappeared presumably because they were hunted to extinction.

Guanacos *(lama guanicoe)* in the Andes of Mendoza Province, Argentina, about 13,000 feet above sea level. (Photo by M. A. Mares.)

The birds of the Monte Desert are also quite different from those of the Sonoran Desert. Andean condors (*Vultur gyrphus*), rare over much of their range, are common in the mountains of northern Argentina. Rheas *(Rhea americana)*, large ostrich-like flightless birds (5 feet tall, weighing 50 pounds or so), are seen often near Andalgalá, especially in areas where they have not been regularly hunted. Rheas feed on flowers and insects and are frequently seen racing through the mesquite forests along the dry washes. Darwin's rhea *(Pterocnemia pennata)* occurs at elevations above 13,000 feet in the high desert of northwestern Argentina. The species becomes fairly common in the scrub desert of the far southern Monte and adjacent Patagonia. Only a few species of birds are shared between the North and South American deserts, and even at higher taxonomic levels, such as families, there are great differences between the faunas.

My first task was to determine the composition of the mammal fauna of the Monte. Before I could study ecology and determine the diversity and abundance of the small mammals, I had to find out which ones were there. I began trapping in the desert. Each night I set out traps with the greatest expectations, but each morning they were empty. I was surprised because in North American deserts, half the traps set out at night might be filled with eight or more species of mammals. I wrote despairingly in my field notes: "Nothing—absolutely nothing. I can't understand why I can't catch any rodents." My inability to trap small mammals would continue for months, during which time I questioned everything from my initial hypotheses to my abilities in the field.

I do not wish to give the impression that the Monte lacked mammals. My trapping was painfully unsuccessful, but some species were only too apparent. Along the mesquite forests lining the dry washes there were many desert cavies *(Microcavia australis)*, tailless, herbivorous guinea pigs that look like North American ground squirrels. The half-pound rodents live in colonies and feed on green vegetation. I often saw them high above the ground eating mesquite leaves, although I found it odd that a roly-poly rodent would have no tail to use in keeping its balance while climbing trees. The animals appeared uncomfortable in the trees and had a difficult time descending if they perceived a threat. Nevertheless, the mesquite trees provided an abundance of green leaves, and the cavies were efficient, if ungainly, climbers.

A desert cavy *(Microcavia australis)*. (Photo by Jon Rood; Mammal Slide Library, American Society of Mammalogists.)

The Monte also has Patagonian "hares," or maras *(Dolichotis australis)*. They are not true hares, which belong to the order that contains rabbits, but are enormous rodents, weighing up to 30 pounds, that race through the desert scrub at more than 30 miles per hour—faster than any other rodent in the world. They occur in pairs or small family groups consisting of the parents and one or two young. They are shy, probably because they are heavily hunted, not only for their pelts—which are made into wall hangings, bedspreads, and rugs—but also for their meat, which is delicious.

There are several predators in the Monte. Foxes *(Pseudalopex griseus)* are seen regularly, although they are also heavily hunted for their pelts, which are sold to furriers in Buenos Aires. Commercial hunting was a major conservation problem when I arrived. Each fox pelt was worth about a dollar. Pelts of wild cats were worth almost five dollars. The hunters, like many peo-

The Patagonian "hare," or mara *(Dolichotis australis)*, the fastest rodent in the world. (Photo by M. A. Rosenthal; Mammal Slide Library, American Society of Mammalogists.)

ple in the interior, had almost no income in the form of currency. They raised goats for food and bartered for other necessities, so any money obtained from pelts was a windfall. The impoverished hunters dedicated most of their waking hours to trapping fur-bearing mammals. There were probably thousands of hunters scattered throughout Argentina in the 1970s, and they were very good at their work.

In the 1980s, my student Ricardo (Dickie) Ojeda and I did a study that showed that 3.6 million fox pelts had been legally exported from Argentina from 1976 through 1979—900,000 per year! Skins from Argentina's mammals fed the vast fur industries of Europe and the United States. The astronomical numbers of exports that we documented for many species led to a renewed interest in controlling commercial hunting. The reports we published on the decimation of native wildlife may have impelled Argentina to initiate regulations controlling wildlife exports, eventually making it illegal to hunt foxes and other mammals throughout the country. In the decades since these controls were implemented, the populations of foxes and other fur-bearing animals, which had become dangerously low, have rebounded in the Monte Desert and they are once again common.

The puma (Puma concolor) is present in the Monte, although it is not common except along the upper slopes of the desert or in areas where desert and dense thorn scrub meet. Since there are few large natural prey left, pumas prey mainly on domestic animals, such as goats, donkeys, and horses. The puma is the same species as the North American mountain lion, and has the largest geographic distribution of any mammal in the New World, extending from far northern Canada to the southernmost tip of South America, a distance of 9,000 miles.

The Monte has two species of weasels, a larger one, the lesser grison (Galictis cuja), at 2 pounds about the size of a small mink, and a smaller one, the Patagonian weasel (Lyncodon patagonicus), which is the size of a half-pound North American long-tailed weasel. These nocturnal animals are seldom seen; they feed on birds, small mammals, lizards, and snakes. The lesser grison was domesticated and used to hunt chinchillas (Chinchilla laniger) in the high Andes, so that the fantastically valuable and luxuriant pelts of these rodents could be used in the fur trade. The chinchillas were hounded almost to extinction. For many years they were thought to be extinct, although recently some isolated populations have been discovered in Chile and Argentina. Now that the animals are protected, and with populations being raised in captivity for use in the fur trade, the species is making a gradual recovery in the wild.

Armadillos are another unusual group of mammals common in the Monte. I frequently observed them scurrying about during daylight hours on cool, winter days or during the night in the hot summer. Their deep burrows—they dig at least one each day—dot the desert floor. There are hundreds, if not thousands, of these burrows in a few miles of desert scrub. Each probably lasts for decades given the low rainfall, so the numbers of burrows grossly overestimates the number of armadillos. Only one species, the screaming armadillo *(Chaetophractus vellerosus)*, is found near Andalgalá. The name comes from the scream the adults emit, which sounds like the shriek of an animal being killed. It is a startling sound when it splits the darkness of a still desert night. This armadillo is smaller and more flattened than its North American relative. It is a consummate burrower and usually attempts to dig rapidly into the soil to escape any predator.

Two other armadillos occur in the Monte. One is the pichi *(Zaedyus pichiy)*, which looks much like the screaming armadillo, but is smaller and hairier. The pichi is omnivorous, feeding on everything from plants to insects to young mammals and carrion. The third armadillo is the remarkably tiny and rare pink fairy armadillo *(Chlamyphorus truncatus)*, perhaps one of the most unusual mammals in the world. It is very small and beautiful, less than 6 inches long and weighing less than 4 ounces. It is covered in brilliant white silky fur and has a leathery, cream-colored shell that extends over the top of the body, attaching fore and aft at only two points. Curiously, it has dense, white fur *under* the shell. Its eyes are tiny and the feet, compared with the size of the body, are enormous with huge claws. These armadillos live in sandy areas, where they "swim" through the sand looking for invertebrates such as beetles and worms.

The larger armadillos, like the predators and larger mammals, are heavily hunted, usually with dogs. They are considered a culinary treat and are killed and roasted in their shells on a bed of coals. The shell is also used to make a musical instrument called a *charango*, similar to a very small guitar.

The pichy armadillo *(Zaedyus pichy)*, a digging machine with a bony head shield, long claws, powerful legs, and small ears and eyes. (Photo by M. A. Mares.)

It is primarily a Bolivian instrument with ten fine metal strings that create a delicate sound unique among musical instruments. It is a difficult instrument to play and mastering it requires great dexterity and extensive training. The Toba and Mataco Indians of the Great Thorn Forest also use armadillo shells in an unusual manner. They fashion the hollowed-out bony shell that covers the tail into a container to carry fine grass clippings and other combustible material and then apply a spark from a flint to it to light a fire, in effect using an armadillo-tail cigarette lighter.

Tucu-tucos *(Ctenomys)* are burrowing caviomorph rodents that look and act like North American gophers, except that they sing, an unusual behavior for any rodent, much less a species specialized for life underground. Like gophers, tucu-tucos have small eyes and ears that do not fill with sand when they dig, tubelike stocky bodies, and powerful legs and arms with long claws on large feet. Gophers dig with the forefeet, then use the face and forefeet (the latter covered with stiff hairs to make them even more shovel-like) to push the dirt out of their burrows and form the mounds that indicate their presence. Tucu-tucos also use the forefeet for digging, but they use the hind feet to push the dirt out of the burrow system, backing out of the burrow rump first as they form the mounds. Their hind feet, rather than their fore-feet, are covered with stiff bristles that assist in moving the soil.

The mound of a burrowing tucu-tuco (Ctenomys) in the Monte Desert. (Photo by M. A. Mares.)

The name tucu-tuco is onomatopoeic and reflects the call that the animals make, a call well described by Darwin: "This animal is universally known by a very peculiar noise which it makes when beneath the ground. A person, the first time he hears it, is much surprised; for it is not easy to tell whence it comes, nor is it possible to guess what kind of creature utters it. The noise consists in a short, but not rough, nasal grunt, which is monotonously repeated about four times in quick succession."

Darwin was describing a species different from the one that occurs near Andalgalá. As I studied the animals across much of their range, I found the calls made by different species could differ from one place to another (a fact also noted by Darwin). The animals near Andalgalá sit in their burrow openings and give a deep, low-frequency thumping call with the accent on the first syllable, "TUcu-TUco, TUcu-TUco," which sounds like a small drum being played below the sparse desert soil.

My wife and I sat in the desert in the midst of a tucu-tuco colony noting the calls over twenty-four-hour periods. The animals call both day and night, with a peak in frequency in the morning soon after sunrise, a second peak late in the day, and a third shortly after midnight. The calls come in waves, so that you first hear animals thumping in the distance, then nearby animals call, and then you hear animals that are farther away begin calling. The distant wave of sound approaches slowly until gradually the punctuated thumping of invisible tucu-tucos is all around; then the deep drumming rolls away across the desert, disappearing into the distance. The reason tucu-tocos call is not clear. It could be a territorial call and it could be that only males call, but nobody knows for sure.

A tucu-tuco *(Ctenomys fulvus)*. Note the tiny ears and eyes and large hind feet of this burrowing specialist. (Photo by G. C. Hickman; Mammal Slide Library, American Society of Mammalogists.)

Tucu-tucos are among the few animals that can live on creosote bush *(Larrea cuneifolia)*, the biochemically rich and toxic plant whose extracts were used for years to coat railroad ties, telephone poles, and fence posts to keep them from decomposing or being eaten by pests. The plant produces hundreds of toxic chemical compounds. I found places where the tucos had opened their burrows under a creosote bush and proceeded to cut down the entire plant, thus decreasing the plant cover of the desert. Few other perennial plants were present, and creosote bush was by far the dominant plant of the area that supported a high density of the rodents.

The ability to forage on a plant that is a veritable toxic waste factory is evidence of a long association between tucu-tucos and creosote bush. The genus *Larrea* developed in South America. Only a single species of creosote bush *(Larrea tridentata)* colonized North America within the last 50,000 years. Since then three distinct genetic races have evolved, one in each of the three major hot deserts (one has two sets of chromosomes and is a diploid, one is a triploid, and the third is a tetraploid). There are five different species of *Larrea* in the South American deserts, a number which supports the idea that the genus originated there. Presumably, tucu-tucos have been associated with the southern desert for a long period of time and thus the rodents have had time to overcome the chemical defenses of the plants.

The tucu-tuco of Andalgalá is also associated with the Mendoza heliotrope *(Heliotropium mendocina)*, a small perennial herb that occurs in bunches on the study site just west of Andalgalá where I was investigating the ecology of tucu-tucos. The clusters of heliotrope plants are associated with the mounds left by the animals on the surface as a result of their digging activities. The plant produces (and develops from) small, tasty (at least to me) underground tubers. The tubers are eaten by the tucos, but are also shoveled out in the soil as it is removed from the burrows when the animals dig their tunnels. The tubers are deposited in the thousands of mounds that dot the desert. After summer rains, the *Heliotropium* sprouts from the tubers that were swept into the mounds, which prove to be ideal places for plant growth, given their high level of aeration and the abundance of plant and animal waste matter. In effect the *Ctenomys* plant next year's food crop in the many mounds that they build. Thus these animals are little farmers like the heteromyid rodents of North America, but they plant tiny potato-like tubers instead of seeds.

Mendoza heliotropes blooming in the tucu-tuco study area in the Monte Desert near Andalgalá, Catamarca Province, Argentina. Each cluster of flowers grows on a mound deposited by the tucu-tucos. (Photo by M. A. Mares.)

Tucu-tucos are extraordinary excavators. Their tunnels are dug more than a foot below the surface—a depth at which winter and summer temperature differences are reduced and a relatively constant temperature is maintained. Burrows may extend for 300 yards or more. They have dozens of openings— small holes that open directly onto the ground surface with no mound and the more typical openings associated with mounds that characterize both gophers and tucu-tucos. The holes that do not have mounds may serve many functions. I know that the animals use these to obtain vegetation, for I watched them open them under plants so they could feed on the plants. I also know that they sing below ground from these holes, for they could not do so from the usually plugged burrows associated with the mounds. The unmarked openings could also function as air vents for the burrow system. There is still a great deal to learn about tucu-tucos for their ecology is complex. In southern Argentina a species was discovered that is colonial, with many animals making up a colony. Social systems vary within the genus, but no one knows why.

Despite my work with the tucu-tucos and observations on a wide variety of other desert mammals, I was unable to catch any sigmodontine rodents so that I could begin answering the major question of my study: How similar are the deserts of Arizona and Argentina? Day after day, week after week, I trapped, but with never an animal to show for it. The most abundant species of mammal was the cavy, but not one entered a trap: cavies seemed to be untrappable.

One night while driving south of Andalgalá I saw a dead viper along the road (vipers are poisonous, nonrattling relatives of rattlesnakes). The Monte has a rattlesnake *(Crotalus durissus)* and a viper *(Bothrops neuwiedi)*; they are the principal venomous desert snakes. I had not yet seen the rattlesnake and this was my first viper. I stopped the car and collected the snake, noting that it had recently eaten. I worked the food item back up its intestinal tract and found a mouse. It was a gerbil mouse, *Eligmodontia typus.* I had never before seen the species, but I recognized it quickly because of certain characteristics of the hind feet that differentiate them from the hind feet of all other rodents on the continent. This was my first indication that there were small mice in the area, if only I could catch them. The only thing I knew for certain about them was that rodents in the Monte Desert were extremely rare, which was the exact opposite of the situation in the Sonoran Desert.

Eventually I found a habitat where rodents and other small mammals were common. The freshwater rivers that flowed from the mountains to the upper parts of the desert valley supported a forest of lush green trees and herbs. In this moist habitat I caught five species of rodents and a small marsupial mouse opossum *(Thylamys pallidior).* The mouse opossum was a tiny mammal (weighing less than an ounce) that was steel gray above and pure white below. It had enormous ears and huge, black eyes, a pointed snout, and a pink nose. It was a beautiful marsupial with a long tail in which it was able to store fat during the winter, increasing its diameter two to three times.

A gerbil mouse *(Eligmodontia)*. (Photo by M. A. Mares.)

The tail was prehensile, and could thus be used to hang from trees or to keep from falling off branches. Mouse opossums spend a good deal of time in trees. They feed on bird's eggs, insects and other invertebrates, and fruit. For the most part, their biology is poorly known. No one has ever studied a desert species in any detail.

The small desert gerbil mouse that I had found inside the viper did not occur in the lush habitats along the river. Clearly, this desert was different from the Sonoran Desert, where ten or more species of small mammals occur in the open desert. In Arizona, if you move from the desert into areas where streams support rich vegetation along the banks, the number of small mammal species actually declines. In Argentina the situation was reversed. Moreover, the two very different groups of rodents in Argentina that had radically different evolutionary histories appeared to have responded to the desert in different ways.

An adult mouse opossum *(Thylamys pallidior)* in the Monte Desert, San Juan Province, Argentina. (Photo by M. A. Mares.)

It was becoming clear that the small mammals of the Monte were not as ecologically similar to those of the northern deserts as I had expected. It is difficult to discard an idea when it is deeply ingrained. Having grown up in a northern desert, having first studied mammals there, and having seen many articles that compared Old World deserts to those of the New World and noted great similarities between their rodent faunas, I was strongly biased toward finding what I expected to find. The faunas should be similar. Rodents should be common in both, and major niche types should be repeated. That is what the literature suggested for other desert comparisons. However, the Monte was not Arizona, and it took many months for me to reformulate my question. Why are the mammals of the Monte Desert different from those of the Sonoran Desert, given the fact that the deserts are so remarkably similar in other aspects?

Kangaroo rats, pocket mice, and field mice of the deserts of North America are famous for needing little or no free water to exist. If fed a diet of dry seeds and no drinking water, many animals can live for years. Hundreds of studies had examined water, salt, and temperature balance in North American desert mammals over the previous three decades, but no similar study had been done for mammals in the Monte. I began a research project on the physiology of Monte species by bringing them into a makeshift laboratory where Lynn and I could control food and water, and where they could exist in stable temperatures. In Mendoza, I encountered two North American amphibian physiologists. They gave me access to a real physiology laboratory with modern equipment, and this allowed me to conduct sophisticated physiological research on many species of small mammals. I wanted to know how South American species compared, physiologically speaking, to species in the northern desert. The two faunas seemed to differ ecologically as far as the types of mammals and their abundances were concerned, but how different could they be physiologically? As Knut Schmidt-Nielsen had shown in his pioneering studies on desert rodent physiology, if there was one area where great similarities had been found among desert rodents from different continents, it was in the area of physiological adaptations to water conservation.

I soon found that Monte mammals could not survive more than a few days without drinking water. One exception to this general rule was the gerbil mouse, which could not survive indefinitely without drinking water, but could drink salty water that was three times more concentrated than sea water. At the same time, the gerbil mouse produced urine that was extraordinarily concentrated. I had expected Monte mammals to be as adept at surviving without free water as mammals in other deserts, but my findings were not supporting this idea. Not only were small mammals rare in the Monte, but they did not seem to be highly adapted to exist in the arid desert.

As I traveled throughout the desert I found that the pattern that I saw in the arid valley where Andalgalá was situated held true for much of the Monte. Caviomorphs were more common in the arid desert than they were in surrounding moist habitats, and sigmodontines, those South American relatives of the North American field mice, were more common in the moist areas. Only the gerbil mouse (*Eligmodontia*) and the gray leaf-eared mouse (*Graomys griseoflavus*) were sigmodontines that regularly inhabited the lowland desert, and the latter was generally restricted to the mesquite forests that grew along dry gullies.

The population density of Monte sigmodontines was also low. It was not uncommon to set 200 traps and capture only a single animal. This was different from the situation in North America, where more than half the traps on any night might capture animals. The ground-squirrel–like cavies were abundant under large mesquite trees along washes, but they were spottily distributed. Even the tucu-tucos, the troglodytic tenors of the rodent world, were abundant only in small patches of habitat—they did not occur over most of the desert.

Further south in the desert there were caviomorphs that seemed to thrive in arid habitats: the Chacoan cavy *(Pediolagus salinicola)*, the plains vizcacha *(Lagostomus maximus)*, the plains vizcacha rat *(Tympanoctomys barrerae)*, and the brush-tailed vizcacha rat *(Octomys mimax)*. None had been studied to any degree, particularly ecologically or physiologically. Indeed, the plains vizcacha rat was known only from a partial specimen. It had not been collected for forty years. Similarly, the brush-tailed vizcacha rat was unknown ecologically, and few specimens existed in museums. Nevertheless, there was a certain diversity of caviomorph rodents, even if they were not abundant. This was clearly a desert of caviomorphs not sigmodontines, and the caviomorphs did not seem to have followed the same evolutionary pathways as other desert rodents of the world.

Different Actors, Different Scripts

Convergence has occurred at all taxonomic levels . . . and to all degrees, from so great as to have been formerly mistaken for close genetic affinity by able taxonomists . . . to virtually none . . . Some adaptations simply appeared in some places and not in others where they would nevertheless be quite possible as far as the environment is concerned . . . In other cases even identical aspects of environments have been parceled out into quite different niches, with no convergence of the animals adapted to them.

George Gaylord Simpson, *The Geography of Evolution*, 1965

G. Evelyn Hutchinson, one of the leading ecologists of the twentieth century, wrote a book in 1965 called *The Ecological Theater and the Evolutionary Play*. The flora and fauna of the Monte and Sonoran deserts could be considered actors in an evolutionary play as the species and ecosystems developed over time. I was certain initially that the actors would be different between the two deserts, although I expected the evolutionary play to have the same script and ending. Clearly, even cursory information made clear that few species or genera were shared between the amphitropical deserts, so there was good reason to expect that the actors would be different. The geology and climate were fairly similar, however, and thus the stage on which the play was performed was the same.

Any investigation of convergence must deal with adaptation at the level of species, even though the higher-order relationships of the organisms must also be considered. It is also at the level of species (and their individual members) that deserts are colonized. A desert does not form from nothing, however. If the climate had been wetter in the past and the area was subjected to climate change, then the transition to desert may have occurred rather swiftly, perhaps over a span of thousands of years. If a barrier such as a mountain range arises to block moisture-laden winds, change may come over millions of years. In either case some type of mammal lived in the area as it began to change from a wetter to a drier habitat. The species that were there when the climate changed either had to be preadapted to survive in the new climate or had to face one of several tough choices: adapt through evolutionary change, leave the region, or die.

Those populations that develop genetic changes that predispose them to survive in an increasingly arid habitat will be the animals that leave more offspring to inhabit the region in future generations. Over time this process leads to the formation of desert specialists, animals that are highly adapted to live in the driest desert. In order for such species to evolve, the time scale of change must be in the millions of years, for the number of changes that must occur in the anatomy, physiology, behavior, and ecology of organisms specialized for desert life is very great.

Deserts are challenging habitats for mammals, which must maintain a high body temperature, be active year-round because of a high metabolic rate, and drink water to cool the body and maintain osmotic balance. Most mammals are ill suited for life in a hot, dry environment. In a desert, green vegetation may be sparse, and those plants that manage to survive may be loaded with toxins (think of creosote bush, cacti, or saltbush). Moreover, plants may be green only for short periods of time, and then the mammals must find an alternative food source or develop the ability to store food.

Annual plants have been especially effective at colonizing deserts. These plants germinate, grow, flower, and fruit within a week or two after a rain, then spend years or even decades as seeds waiting for the next rain. Thus for much of the year there may be little more for a small desert mammal to eat than seeds. On the positive side, there are always seeds in a desert.

If a species can specialize on seeds it assures itself a food resource in the driest habitats. But existing on a diet limited to seeds is not easy, and a life dedicated to finding, storing, and eating seeds requires many adaptations of teeth, feet, eyes, ears, behavior, physiology, burrow maintenance, and more. Feeding on seeds also sets an upper limit on body size of about a quarter of a pound, since larger-bodied mammals cannot gather enough small seeds in the desert to survive. A larger-bodied mammal would expend more energy collecting seeds than it would obtain from the food value contained in the seeds.

It is through the many limitations imposed on small mammals in a desert by the climate, soils, and vegetation that natural selection molded the adaptations of the ancestral species that first encountered arid habitats. Their progeny became the species that we now recognize as desert specialists. Comparative studies of desert mammals suggest that there are only so many major lifestyles, or niche types, that are successful strategies for small mammals to follow if they are to inhabit a desert. The North American bipedal desert rodents provide a good example of a major niche type in the desert.

Most early ecological research on desert rodents was done in the United States, where the bipedal rodents are either kangaroo rats or kangaroo mice. Almost all species eat seeds and the logical connection was made that the animals had become bipedal because that body shape facilitated foraging for seeds. Thus the forelimbs became small in response to the need to gather seeds, and the animals were able to cover larger distances searching for seeds by virtue of their unusual method of locomotion. It was a plausible explanation, but it was not correct. At the time this hypothesis first appeared few ecological studies had been conducted on the Old World jerboas, another important group of bipedal desert rodents. It was assumed that the Old World animals functioned like the kangaroo rats they so strongly resembled.

Not all kangaroo rats eat seeds, however, with the chisel-toothed kangaroo rat being a notable exception, as we have seen. Moreover, most of the seed-eating rodents of North America's deserts are not kangaroo rats, but are pocket mice, smaller relatives of kangaroo rats that are not bipedal. Finally, many of the field mice of North American deserts also include seeds as a major component of their diet, especially during parts of the year when other foods are scarce.

Nonetheless, through the power of repetition rather than through scientific research, all bipedal rodents were assumed to be seed eaters. If rodents specialized for a life in an arid habitat they would very likely eat seeds and hop about on their hind legs. Kangaroo rats behaved this way and species that resembled kangaroo rats, it was said, must be similar. Thus Old World desert rodents that looked like kangaroo rats were automatically assumed to be ecological analogues of the North American species. This similarity became a "classic" example of convergent evolution.

When I traveled to Argentina in 1970 for the first time I was steeped in such thinking. It was only after being exposed to a fauna that was alien to me and after learning about the natural history of the South American species that I began to question whether the prevailing wisdom was correct. In Argentina there were no bipedal mammals at all, although there were many species of mammals living in the desert, some of which appeared to have been associated with aridity for a long time. I assumed that most of the small desert mammals that belonged to the field mice group (which had entered South America from North America) had reached the desert too recently to develop the high degree of specialization that characterized North American desert species. It seemed to me that bipedal seed eaters ought to live in the Monte Desert, since it was an old desert and was similar in most respects to the Sonoran Desert, but there did not appear to be any.

Adaptations are fashioned as environmental challenges are met and overcome. Given the rich diversity of the genetic pool of mammals, and the force of mutational change and natural selection, you might expect that environmental challenges would be overcome in similar ways. If a small mammal develops the ability to exist without free water, for example—something that in principle seems difficult to accomplish—surely similar anatomical and physiological specializations will be required of the kidneys and associated organs. If the animals in two deserts are only distantly related, the similarities will be the result of convergent evolution—the molding of adaptations along similar paths toward similar endpoints to overcome similar evolutionary challenges. Rodents in other deserts that had been studied looked and apparently functioned much like the well-studied and paradigmatic North American species. They appeared to be strongly convergent on the

North American species. Why would the story of mammals in the Monte be any different?

Nonetheless, my work was showing not only that the two deserts that acted as the evolutionary stages had different actors, but that the actors appeared to be following different scripts. As I examined the deserts in greater detail, I found that many of their organisms were very different in the ways in which they managed to exist in the desert. The evolutionary play seemed to have been written by a different author on each continent, working from the same basic premise but giving the plot different twists.

An organism evolves within the context of other species—whether animal or plant—and the mammals of the Monte not only had developed within a milieu unique in this regard, but had done so over periods of time radically different from those that characterized the development of the North American fauna. Some groups in the Monte (the armadillos, marsupials, and caviomorph rodents) had been present since the desert formed. The sigmodontine field mice and several other groups (camels, deer, and carnivores) arrived from North America thirty million years later, finally reaching the southern desert millions of years after it had developed. The stages may have looked similar, but the actors and the plays were indeed distinct.

Small mammals in the desert of Arizona were abundant and diverse, but I was not catching many mammals in the Monte. I began to wonder whether habitats that were wetter would support more species. Could it be that mammals were just uncommon everywhere in Argentina's Northwest, or was it only in the deserts that they were so rare? There were no data on mammal population ecology in the entire region, in either wet or dry habitats, so I began to travel through both the arid and the moist areas of western Argentina to familiarize myself with the fauna of the various habitats.

On an early trip to explore new areas my wife and I entered an almost pristine forest along the Bolivian border in Salta Province, where woodcutters were just beginning to fell trees. It was the first time the forest had been cut, and dirt roads were just being built. Rainfall in the area exceeded 80 inches per year. Bats were plentiful, as were carnivores and field mice. We walked carefully because jaguars, the most impressive cats in the Western Hemisphere, had been seen in the area. A jaguar in the neighboring province had killed a man only a few months earlier. We also saw a jaguarundi (*Herpailurus yagouaroundi*)—a svelte, sinuous black feline about the size of a house cat. It growled at us from a large tree branch as we walked under it.

It was a glorious place to work. Flocks of parrots chattered overhead and toucans flew by, seemingly pulled along by their enormous and colorful bills. Each day we swam in the river and drank its water, for there were no human habitations above us. But paradise can have its negative side. For us it was the insects and parasites. Ants, bees, and wasps assaulted us by day and mosquitoes by night. We did tick patrol every night to remove the tiny parasites, most barely larger than a speck of dust. Generally we would pick off 200–300 ticks after an hour of grooming like a forest primate.

We caught hundreds of animals of a variety of species in just a few days and most were relatives of the field mice. With the exception of one species of cavy, an agouti, and one species of burrowing tucu-tuco (this one apparently mute, for I never heard it call), caviomorphs were uncommon in the area. But marsupials were common, particularly a beautiful mouse opossum (*Thylamys venustus*)—a tiny, 1-ounce opossum that is a rich brownish gray color above and yellowish gold below, with a long tail, large black eyes, and a delicate pointed snout. The fauna of the tropical forest was entirely different from that of the Monte, even though the Monte Desert was less than 100 miles away as the condor flies.

The field mice of this forest included a shrewlike burrowing mouse (*Oxymycterus paramensis*), a long-snouted rodent that forages under the litter of the forest floor and feeds on invertebrates. There were also several species of grass mice (genus *Akodon*), very furry, dark little rodents that look much like North American voles, but unlike the vegetation-eating voles, feed mainly on invertebrates.

On those first trips to the forests of the Northwest we caught a lot of small mammals, some of which we could identify and many to which we were unable to attach a name. There was no scarcity of mammals in northwestern

Tropical montane forest near Calilegua, Jujuy Province, Argentina. This habitat extends along the Andes of South America through Salta Province, Argentina, and is rich in mammals, including monkeys, tapirs, and jaguars. (Photo by M. A. Mares.)

Argentina. The scarcity was in the Monte Desert. In 1971 I was unaware of the great complexity of the mammals of northwestern Argentina, although I was becoming increasingly uncomfortable with my inability to identify many species. Sometimes I could tell that I was seeing several species, but I was unable to attach names to them. Clearly, without adequate identifications, I would find it impossible to do field ecology, which was one of my goals. Nevertheless, as we worked through the forests we found the same pattern of species occurrences that we had observed along the Bolivian border: the ubiquitous South American field mice were abundant and the caviomorphs were less common. In order to determine whether this pattern held in nonforested areas, we climbed into the high pre-Andean grasslands, above 10,000 feet in elevation.

We sampled at El Infiernillo in Tucumán Province. The area was a dense bunchgrass prairie set high in the mountains. It was only slightly affected by human activity. It lacked fences and there was only light grazing from horses and other livestock. In that mountain grassland we discovered an extraordinary concentration of small mammals. We caught more animals in one night than we had caught in six months of sampling in the Monte and trap success exceeded 100 percent. Some traps held two or three animals. Perhaps not surprisingly, the mix of species was different from what we had found in other areas and many species continued to be difficult or impossible to identify.

The vesper mice (*Calomys*) that we had found in the lowland forest were also present in the grasslands. These rodents are important to humans because they carry some deadly viruses, including Machupo virus, which killed hundreds of people in Bolivia in 1962 and has flared up several times since. Machupo, carried by the Bolivian large vesper mouse, *Calomys callosus*, a close relative of *Calomys venustus* (for decades they were thought to be the same species), is an arenavirus, a relative of hantaviruses, and causes a frightening disease that kills quickly by inducing hemorrhagic meningitis. *Calomys* also harbors the virus responsible for Argentine hemorrhagic fever, another deadly disease that kills by causing bleeding from the bowels, kidneys, eyes, and nose.

Like the lowland forests, the high grasslands also supported several differ-

Dry bunchgrass prairie at El Infiernillo, Tucumán Province, Argentina. Before the habitat was altered by overgrazing and agriculture, this area supported extremely dense and diverse populations of small mammals. (Photo by M. A. Mares.)

ent grass mice, a species of leaf-eared mouse *(Phyllotis osilae)*, an unusual grass-eating rodent called *Andinomys edax* that looks like a North American woodrat, and a lovely herbivorous rodent called the bunny rat *(Reithrodon auritus)*. Caviomorphs were represented by the common yellow-toothed cavy *(Galea musteloides)* that we had found in forest clearings and another tucu-tuco *(Ctenomys knightii)* a large, chocolate-brown animal five times the size of the Monte species that emitted an odd whistling call. The high grassland was clearly the domain of the field mice, not the guinea pigs and their allies. There were two major actors on the evolutionary stage in Argentina and each had responded differently to the habitats that had developed over time. Each had been successful in colonizing the valleys and mountains of northwest Argentina, but each had done so in a different way. Caviomorphs ruled the lowland desert and sigmodontine field mice ruled the wetter habitats. Why did these two groups show such different patterns? Also, were small mammals in the isolated Bolsón de Pipanaco, the site selected for the IBP research, unusually scarce? Could other parts of the desert harbor more species or greater abundances of rodents? Perhaps I was based in the most mammal-poor zone of the Monte and was erroneously inferring scarcity of mammals for the entire desert. I decided to explore the rest of the Monte Desert.

The Monte is a beautiful desert with vistas of distant snow-covered peaks visible across shimmering, olive-green creosote flats. The mountains rimming the desert are devoid of trees, their vegetation consisting of cacti, bromeliads, and shrubs. The bromeliads often provide a dense ground cover between cacti and rocks, giving the arid slopes a soft, almost upholstered appearance, or painting the arid hills with surprising and welcome splashes of color when yellow flowers appear on the ends of the tough, woody stalks. The ground bromeliads resemble the yuccas of the American Southwest, but have fishhook spines on stiff green leaves that grow from a basal cluster.

I came to know the Monte well, to respect it, and to love its beauty. North of Andalgalá the elevation of the desert increased through a series of isolated valleys until it terminated near Cachi in Salta Province 8,000 feet above sea level. As you move northward through the valleys of Santa María and Cafayate, the desert changes. Steep, dissected slopes carved sharply by rain and wind display the underlying strata in a riot of multicolored tones of black, gray, yellow, red, and brown. At the base of the Calchaquies Valley a forest of green mesquite, acacia, paloverde, and *Bulnesia* trees outlines a ribbon of clear water. *Bulnesia* is an unusual tree that is almost leafless, photosynthesizing in gray-green stems given their muted color by a thin coating of wax that reduces water loss. In summer during the rainy season all of the various species of trees bloomed at about the same time and produced bright yellow flowers that laid a broad band of gold on the valley floor.

As in the desert near Andalgalá, I found the small gerbil mouse *(Eligmodontia typus)* in the desert valleys of the Monte (the genus continued into Bolivia and Peru). Supposedly only one species existed in lowland Argentina, but I was to learn that gerbil mice are another relatively complex genus containing eight or more species. During my first two years of fieldwork, however, I believed there was only a single species. Indeed, the more I traveled and examined animals in the many habitats of western Argentina, the more I came to realize how ignorant I was of their taxonomy. It was becoming increasingly difficult to identify the specimens. I simply did not know where species limits began or ended, and the published literature was of little help. A discussion of the two complex genera encountered thus far—one from the forest and one from the desert—is warranted, because it illustrates the importance of fundamental data on taxonomy for all higher studies.

The first group is the grass mice, genus *Akodon*. This may be the most complex genus of mammals in the world, and in the wet forests of northwestern Argentina my traps were filled with them. At one time there were more than ninety species named. Splitting the genus into several genera and lumping species together reduced that number, but recent research suggests that the initial complexity of the genus *Akodon* was probably closer to the truth. Species of *Akodon* occur across most of South America, from areas more than 16,000 feet high to areas at sea level and from Venezuela to southern Argentina. They are found in all major habitats. Moreover, they all look very much alike, which makes them a nightmare for field biologists.

Their external similarity makes mark-and-release field studies, where organisms are caught, identified, marked, released, and caught again—the bedrock of ecological research—all but impossible because you cannot be certain of the identifications. If you cannot distinguish among similar species, you cannot study them in any meaningful way. Characteristics of one can be confused with those of another, leading to erroneous inferences. This illustrates the pivotal role played by taxonomists—those who describe species—and systematists—those who show how species, genera, and higher taxonomic categories are related. These scientists provide the foundation for ecological, behavioral, physiological, and evolutionary research.

Biogeographic evidence suggests that grass mice radiated (developed many species over a short span of time) in the Andes of Bolivia and northwestern Argentina—at least that is where most species occur today. After thirty years of studying these animals my colleagues and I can now, in most cases, attach a name to a specimen. But each time we trap in a new, more isolated locality, we encounter animals that we cannot readily identify. In 1997, for example, we collected a tiny, yet fully adult, *Akodon*. It was the smallest grass mouse we had ever seen and proved to be a species new to science. We named it the diminutive grass mouse, *Akodon aliquantulus* (Latin for "so small"). Curiously, we found it on an isolated grassy mountaintop 20 miles from the city of Tucumán in Argentina's smallest and most densely populated province. In 1991 Rubén Barquez, Dickie Ojeda, and I had published a field guide to the mammals of the province, and felt we had a good understanding of its mammals, but the diminutive grass mouse was leading its existence within sight of the capital city of San Miguel, a major metropolis. It was there when the Incas crossed Tucumán while traveling from Peru to their southernmost outposts in Mendoza Province. It remained anonymous until a few traps were set in an isolated grassland. Now it finally has a name. Of its ecology, we know nothing.

The second complex group is the gerbil mice, genus *Eligmodontia*. In 1990, working in thorn scrub just east of the Monte Desert near the town of Chumbicha, Catamarca Province, my colleagues and I captured a species of *Eligmodontia* that had not been seen in the wild for over seventy years.

Many if not most of South America's mammals were named by the greatest mammal taxonomist who ever lived, Oldfield Thomas of the British Museum of Natural History in London. Thomas published more than eleven hundred papers and named thousands of genera, species, and subspecies of mammals from throughout the world, but especially from South America. Among the species he named was *Eligmodontia marica*, a small gerbil mouse collected in 1918 near the town of Chumbicha, Catamarca, by Emilio Budin, one of Thomas's collectors.

Thomas was not a field man, but he understood the value of those who went to the field to collect and study the animals that occur in the poorly explored parts of the world. He also knew the value of museums and recognized that his own career was coming to an end. When Thomas named the lovely little mouse *Eligmodontia marica*, it was his 2,000th type specimen, and he named it in honor of his wife, Mary. In the Afterword to the paper Thomas wrote about this mouse, he said:

> As an indication of the extent to which our British National Museum has participated in the general advance in the systematic knowledge of Mammalia, and the corresponding accumulation of typical [that is, type] specimens, I may perhaps be permitted to record that, so far as I am able to calculate, this is the two-thousandth mammal to which, as the official mammalogist of the Museum, I have had occasion to give a name. And many hundreds more have been described and named by other workers. The vastness of the collection—especially of types—indicated by these figures is due mainly to the patriotism of our countrymen all over the world, many of whom have been proud and pleased to contribute to their National Museum merely because it is the National Museum, without pay or return, and often in climates where mere existence is a burden.
>
> Having possessed for forty years the great privilege of working on this wonderful collection, I feel I cannot too strongly express my appreciation of the generosity and public spirit shown by its many contributors—whether those who at home have provided funds for making expeditions, or abroad have made collections to be added to the National treasures. My own share in the work, carried on as it has been under the most favourable conditions, has been a continuous pleasure. And in appreciation of one important element in this pleasure, the sympathetic and ever-ready help of my wife, I have given to this attractive little animal the above specific name.

Thomas, who eventually would supply the names for the genus, species, or subspecies of twenty-nine hundred mammal taxa, was very close to his wife, who worked as his research assistant. Shortly after her death, Oldfield Thomas committed suicide, putting an end to an extraordinary career, one that has never been equaled. He was seventy-one. According to his biographer, John Hill, "the deliberate termination of his life by his own hand in June, 1929 came as a great shock to his many friends and admirers, although no surprise to those close to him: the note that he left revealed his inner stresses and made it plain that he could no longer face life alone without his partner."

When the late Philip Hershkovitz of Chicago's Field Museum revised the genus *Eligmodontia* in 1962 he concluded—after what can only be termed a cursory examination of the group—that there existed only one lowland species. He was at that time what is called a "lumper" in taxonomy, one who gathers together animals that have been placed in several species and attempts to simplify the situation by combining them into a single species. The problem with lumpers is that if they combine several species into one, they do a disservice by confusing field researchers, who may come to believe that very different animals in the wild represent only a single species.

Defining the limits of a species is difficult, especially if there are only a few specimens available that represent the range of variation of the animals throughout their geographic area or if there is little information on the biology of the animals in the field. When Hershkovitz revised the genus *Eligmodontia* as part of a much larger study, he lacked knowledge of both the range of variation and the biology of the animals in nature; he mainly examined the relatively few specimens that were available to him in Chicago. Thomas had named three species of *Eligmodontia* and several subspecies, but Hershkovitz had never examined the type specimens. When I finally had an opportunity to travel to London to examine the type specimens of *Eligmodontia*, after I had had extensive experience with the various gerbil mice in the field, it became clear that Thomas was correct.

The genus clearly included many species, not just one, and the most distinctive was *Eligmodontia marica*, the tiny, delicate, blond mouse from one of the driest parts of the Chacoan thorn forest that Thomas named for his wife. Thomas included in his paper on this mouse Budin's description:

"This pretty mouse has been the one which has most pleased and interested me of all the rodents." I had worked in Catamarca Province for two years, but had spent no time in the eastern parts of the province, where semiarid Chacoan thorn scrub, rather than arid desert, forms the predominant habitat. Thus I had never had any reason to search for the elusive and tiny gerbil mouse.

As I began to tease apart the species that make up the genus *Eligmodontia* it became clear that Hershkovitz ignored the type specimens at his peril, for anyone could have seen how different this lovely little gerbil mouse was from the rest of the group. Only four specimens of *Eligmodontia marica* existed, and fieldworkers had not encountered the species for over seventy years. Indeed, Budin was the only person who had ever caught one, and that was in 1918, but I was sure that I could catch one by returning to the type locality. I decided to go to Chumbicha.

I knew that Budin had traveled on trains and by horseback. As my colleagues and I arrived in Chumbicha, a tiny town south of the provincial capital of Catamarca, I sought out the old railroad station, which was no longer in operation. The town had changed greatly over the decades, and now had paved roads and newer buildings, and lay along a major highway. Still, if Budin had arrived in town on the train, where would he have set his traps? Where would I set the traps? That should tell me where Budin had set his traps.

A patch of heavily disturbed arid scrubland lay within a mile of the old railroad station. We set our traps in the area even though it hardly looked promising, since cattle, goats, and other domestic animals were foraging there. Nevertheless, Catamarca Province has had domestic animals for more than four centuries, so I thought that the small mammal fauna might still be similar to that which Budin sampled so long ago. We set our traps and in the morning found that we had captured *Eligmodontia marica*. My notes recorded: "The *E. marica* I caught is a lovely, tiny (10 g) [less than half an ounce] mouse—it looks just like the type specimen."

Oldfield Thomas was right. Time and again as I work on rodents that have received almost no attention from mammalogists for as long as a century or more, I find that Oldfield Thomas had an especially discerning eye and an uncanny feel for species boundaries. His contributions to mammalogy are as pertinent today as they were a hundred years ago.

In 1987 in the extremely isolated Monte valley called the Campo Arenal (Field of Sand), my colleagues and I collected a single specimen of gerbil mouse that was obviously a new species. It took us ten years to get back to the site and collect additional specimens. The animals live on large sand dunes, an unusual habit for this group of mice. In 1999 we found yet another un-named species of gerbil mouse living on hard, gravelly soils alongside the dunes. The two new species occur within a few yards of each other, but in very different microhabitats. *Eligmodontia* continues to yield new species as we extend our research through the arid lowlands and high plains of Argentina. It is a complex genus that has developed many species in the arid lands of lowland Argentina.

These two examples of genera that contain many species illustrate why working in South America in the latter part of the twentieth century was like working in North America a century earlier. The mammals of North America are well studied today, and few new species are discovered nowadays. But this has not always been the case. When a good taxonomic foundation is lacking, all areas of scientific research on mammals are affected. This lesson had been learned during the biological exploration of the United States. A situation similar to that surrounding the genus *Akodon* existed in those early days, especially with regard to a large group of rodents that were common throughout North America. These were members of the genus *Peromyscus*, a genus of field mice that includes about fifty species that often occur together in groups of five species or more. The species were extremely difficult to distinguish until the landmark work of Wilfred Osgood, an American mammalogist who tackled the tangled taxonomy in 1909 and finally sorted out the species. He was able to conclude that there were forty-three species of *Peromyscus* in North America, and he also clarified the approximate distribution of each species.

Once the identity of each species became clear, it was possible for the first time to describe details of each one's geographic distribution, habitat selection, habits, behavior, life history, and other attributes. Later work using modern techniques refined our understanding of this complex genus, but Osgood made the greatest contribution, for he drew the map that others could follow. He was the pathfinder. Without a proper taxonomy, other research had been impossible, for how could the ecology of species "A" be described if a researcher were actually examining a mix of species "A," "B," "C," and "D"?

All of this might sound like esoteric academic games except for the fact that a couple of members of this genus happen to be the deer mouse (*Peromyscus maniculatus*) and the white-footed mouse (*Peromyscus leucopus*). In the 1990s, first in the southwestern United States and subsequently in states outside the region, it was found that these two species were hosts for the *sin nombre* hantavirus, a virus as deadly as the more famous Ebola virus of Africa. Shortly after it was established that people were dying in just a few days from a pulmonary virus, the hosts—the two species of *Peromyscus*—were located and measures were taken to alert doctors and the public to places where hantavirus could be expected to be found. This quick action would not have been possible without Osgood's work and the efforts of museum scientists in over a century of collecting. If we had still been uncertain which species was which, how would we have known that it is these two species that are the hosts? How would we have known where they were found and what their associated species were likely to be?

The composite database of mammals of North America that lies within the collections of major museums and the many publications of mammalogists who had been working for over a century told health workers the name of the host. The database also provided data on where the species that carried the virus lived, described their habitats and habits, told investigators where to look for other species that were similar to the infected species and that shared their geographic ranges, and provided a list of the nearest relatives of the infected species. Doctors could be alerted to consider a *sin nombre* infection for a patient with pulmonary distress because *Peromyscus maniculatus* and *Peromyscus leucopus* were common in their area. This database of taxonomic and systematic research may ultimately show how the viruses themselves came to be associated with the rodents, for there is some support for the idea that the rodents and the viruses co-evolved. A solid understanding of the taxonomy of the mice makes viral research easier. By understanding the biology of the rodents, we were able to more readily understand the biology of the deadly virus.

The United States was extremely fortunate to have had such extensive research conducted on its small mammals. After Osgood and the succeeding taxonomists and systematists, it was possible for the ecologists, behaviorists, and physiologists to go to work. Together they provided the fundamental knowledge of the fauna that is required if one is to understand where mammals occur and what they do. This most basic data of the fauna of a country is becoming increasingly vital as habitats disappear, as species slide toward extinction, as new infectious agents arise, and as human health and well-being become increasingly affected by nature. Unfortunately, the situation in the United States, with its solid history in taxonomic research, is not typical of most other countries.

Working in South America, Africa, or Asia, where the taxonomy of the mammals is not well known, one can see the immediate effects of a lack of the foundational data of nature. Ecological studies are few, for the animals cannot be identified. Some published papers report data for misidentified species, rendering the research meaningless. Behavioral and laboratory research is lacking because people do not know the animals exist, or do not know where to find them, cannot identify them if they *do* find them, and know so little about the animals' natural history that more complex ecological studies are not feasible. Moreover, in the realm of public health, new viruses are emerging with regularity, and some virologists think that it is only a matter of time until a highly contagious, deadly quick-acting virus sweeps across the globe. When new and deadly diseases arise, it is much harder, if not impossible, for epidemiologists and infectious disease researchers to determine the host and its geographic distribution if the basic taxonomy has not been done. The ecology of host species will seldom have been studied in most of the world, so dealing with pathogenic outbreaks is made much more difficult. Good systematic information and solid data on the natural history of potential rodent vectors offer a lifeline of hope to virologists and public health specialists, who have to predict the course of viral diseases and develop plans to combat them.

In 1971 and 1972, however, as I surveyed the mammals of the Monte, I knew nothing about the tangled taxonomy of *Akodon* and *Eligmodontia*. I assumed that I was encountering several species of *Akodon*, and indeed I could attach a name to a number of them, but there were clearly more species in the various habitats than the current wisdom suggested. As far as *Eligmodontia* was concerned, I followed Hershkovitz in recognizing only one species in the lowlands.

The Monte is even more arid south of Andalgalá than it is in the north. There are places where annual rainfall is only an inch, making these areas as dry as the Mojave Desert. As in the Mojave, plants in these regions are exceedingly sparse and even cacti are rare. If mammal abundance and diversity were low around Andalgalá, they were even lower farther south. Tucutucos are not found in the most arid habitats, the plant density being too low to support their populations. Trees become exceedingly sparse, so cavies disappear. Only the gerbil mouse is captured occasionally. Rarely, a leaf-eared mouse *(Graomys)* is found. The large deerlike rodent called the mara *(Dolichotis)* is present, but uncommon.

I extended my research southward to the almost vegetation-free Valle de la Luna (Valley of the Moon) in San Juan Province, an aptly named region consisting of naked eroded hills and geologic formations sculpted into weird shapes by wind and water. The area is so dry that plants that grow into large trees in other parts of the desert exist there as tiny shrubs. Paloverde trees that might reach 20 feet in height in the north seldom exceed 1 or 2 feet in height in the Valley of the Moon. The gerbil mouse was the only species I found in this most arid desert until I encountered the brush-tailed vizcacha rat *(Octomys mimax)*.

I found the brush-tailed vizcacha rat to be uncommon and difficult to trap. Burrows and tracks showed that a rodent was present, but I did not know what species was involved until I was finally able to capture a single individual. The rodent looks very much like a North American woodrat. Like the woodrat, it eats cacti, which are not abundant in this hyperarid zone. *Octomys* belongs to a most unusual family of rodents called the Octodontidae. The group is ancient for rodents, with fossil octodontids known over a span of thirty-five million years. They have been associated with the deserts of Argentina since the deserts formed. *Octomys mimax* was highly adapted to life in the desert and existed only in the most arid sections.

I soon learned of another rare desert mammal—a second species of octodontid rodent called the plains vizcacha rat *(Tympanoctomys barrerae)*. The plains vizcacha rat purportedly lived in salt flats, but no specimens had been collected for more than a quarter of a century and only a partial specimen was known. I traveled south to eastern Mendoza Province in 1971 to search for the animal near the type locality, the place from which the type specimen—the actual animal on which the scientific name is based—was captured. I sat in the desert enveloped by a moonless night. There were few mammals in this arid, flat area located far from the Andes. Even the calls of the tucu-tucos and the screams of the armadillos were absent. There were no cities within a hundred miles. The stars were brilliant, but their sparkle did not illuminate the dark plain.

Suddenly I was startled by loud whooshing noises just above my head. I quickly turned my flashlight toward the sky and saw a huge rose-colored wave of flamingoes flowing over me on their way to a distant salt flat. My light startled them as much as they startled me and they began squawking noisily. For a brief moment the black night exploded into a riot of fiery pink against the raven sky. I was again reminded that life in the desert extends well beyond mammals and that this desert, like others, was teeming with life even if I was unaware of its existence.

My traps yielded only a single gerbil mouse, no plains vizcacha rats, and no other mammals. I was unable to search again for the plains vizcacha rat until the mid-1990s, and returned only after Dickie Ojeda, my former student, had collected several animals a few miles from the point where I had seen the flamingoes two decades earlier. Dickie had invited a group of Chilean mammalogists to collect specimens, and the Chileans found that

Tympanoctomys barrerae was a very unusual animal, having the largest number of chromosomes known for a mammal. Later research suggested that the species is a tetraploid, meaning that it has a double complement of chromosomes. Tetraploidy is fairly common in plants and has been shown in a few other vertebrates, but the plains vizcacha rat is the only mammal with this trait. How a double complement of chromosomes might have evolved in a mammal is something that can not be explained with our current understanding of mammalian genetics.

Mapping the burrows of the plains vizcacha rat *(Tympanoctomys barrerae)* in southern Mendoza Province, Argentina. (Photo by M. A. Mares.)

I was finally able to study the ecology of the plains vizcacha rat in detail in 1995 in a salt flat in southern Mendoza Province. In the area where the Chileans worked *Tympanoctomys barrerae* lived in rather small burrows under salt bushes, but the plains vizcacha rats of central Mendoza lived in massive mounds having dozens of openings that led into a complex system of crisscrossing burrows. The mounds rose 3 feet above the surrounding salt flat and were almost 30 feet long. They were larger than those used by any other small rodent. Nothing was known about the biology of the species, but I believed the animals must be colonial, for such a large mound could hardly be maintained by a single rodent smaller than a woodrat.

Unfortunately, my colleagues and I could not capture the animals, despite the use of various types of traps and baits. In desperation I decided that we would dig the animals out of the mound. We plugged the many openings with traps so that if an animal was present it would go into the trap as it raced out of the mound. After all burrow openings were plugged, we began digging away at the enormous quantity of soil. We expected several animals to come out of the mounds as they fled the burrow.

As we dug, we mapped the vegetation and the openings of the burrows. Within each mound, the burrows divided and divided again, with passages on three different levels. It was a complex warren. Eventually, after much of the first mound had been excavated, we heard a trap door slam shut. We had caught our first *Tympanoctomys barrerae*. I was certain that more animals would be leaving the mound, but though we completed the excavation we captured no other animals.

We proceeded to excavate another half dozen mounds, and each time only a single animal inhabited a mound system. Using our newly acquired information about how many animals inhabited a mound, we were able to census the population of the area by counting and mapping active mounds. We learned that the area supported only a single animal for every 3 acres of prime habitat, but the prime habitat was very limited. As we moved away from the salty vegetation that surrounded the salt flat in a narrow band, the number of mounds quickly fell to zero. The animals lived only in the halophytic vegetation immediately surrounding the salt flat, vegetation that extended no more than a hundred yards or so away from the crystalline salt. We had managed to learn something about the natural history of the animals in a rather short time, especially considering that there was no information available on their ecology. The animals were solitary, with a single animal inhabiting each complex mound. Population density was only one animal per 3 acres and their habitat was only the immediate periphery of the salt flat. They foraged on saltbushes, were nocturnal, and were remarkably attached to their mound system, being loath to abandon it even after we began to dig into it.

The plains vizcacha rat *(Tympanoctomys barrerae).* (Photo by M. A. Mares.)

I placed one of the animals in the aquarium that always accompanied us into the field to observe its behavior. This rat is a beautiful rodent with a large head, doleful eyes, a rounded, rather chubby, body, and a thick, bushy tail. Its fur is soft brownish gray, and it has a white belly. The animals are easily handled and do not bite. As I observed the rat I decided to offer it some plants from those few species that occurred near the mounds. We already knew that the animals ate saltbush because parts of the plant were scattered over the mound and in the tunnels. Moreover, saltbush is the dominant plant in the area. If they did not eat saltbush there would be little else for them to eat.

I gave the rat saltbush and it quickly picked up the plant and proceeded to eat it rapidly. As I watched, small bits of plants flew in all directions. Some stuck to the aquarium walls and, after the rodent had finished eating, I collected the tiny white chaff from the glass for closer inspection. The bits consisted of the outer coverings of the saltbush leaves, those specialized cells where the salt crystals are stored. The plains vizcacha rat, like *Dipodomys microps*, the saltbush-eating chisel-toothed kangaroo rat of the Great Basin Desert, was peeling the protective salt covering from the leaves and feeding on the green plant tissues.

How did the animal accomplish this feat? When the chisel-toothed kangaroo rat eats saltbush it carefully scrapes the leaves over its specialized lower incisors in a way that removes the salt-storage cells. It is a laborious process and one that is not too hard to observe. *Tympanoctomys barrerae*, however, had made these tissues fly! I could not imagine how.

I grabbed the rat and examined its mouth very closely. Inside, largely hidden from view, were two toothlike bundles of stiffened hairs on either side of the mouth that resembled two extra upper incisors (rodents have only a single pair of upper incisors). When my colleagues and I examined the structures under the microscope, we found that they were unique among mammals. The bristles were shaped exactly like a tooth and were continually sharpened against the lower incisors, so that they maintained a honed edge. Moreover, the lower incisors had the same shape as those of the North American chisel-toothed kangaroo rat—they were shaped like a chisel.

I again watched the rat eat saltbush. Its face seemed to vibrate just before it began to feed the saltbush into its mouth, where the rapidly moving lower incisors and the vibrating bristle bundles quickly stripped away the salt-filled leaf cells. The mouth appeared to function like a bark stripper in a sawmill. As the salt-filled tissues were sheared away, the green tissue was exposed and eaten by the rodent.

Incredibly, and unexpectedly, we had found a South American rodent that specialized on a diet of saltbush leaves, one of the rarest of adaptations among mammals. This odd animal had evolved a mechanism that was both similar to, and yet distinct from, that employed by the chisel-toothed kangaroo rat of North America. This rodent actually used specialized hairs to help it process its food. Few mammals use hair to assist in gathering food. The most ready examples are the baleen whales, those massive oceanic beasts that have developed filters of modified hair—the baleen—that strain microorganisms from the sea as the whale pumps ocean water through the mouth. In one respect, the little plains vizcacha rat is a whale of rodent.

This degree of adaptive specialization on a single desert plant by *Tympanoctomys barrerae* suggests a long history of interaction between the two species. The bristle bundles appear to have evolved as anatomical adaptations that facilitate removal of the salt covering of the saltbush leaves. No other plants in the area seem to require such a structure. The kangaroo rat of the Great Basin and the vizcacha rat of the Monte have converged on each other in both form and function, even though they do not resemble each other externally. Both species are salt specialists, live in similar habitats, and have chisel-shaped teeth and specialized kidneys—the kidney possesses an exceptionally elongate Henle's loop, which assists in urine concentration, a requirement for processing the high salt loads obtained from the saline plant fluids. The two species' habit of foraging on an unusual desert plant with a striking defense mechanism has meant that they had to overcome the same evolutionary hurdles in more or less the same manner, with the Monte Desert species developing the unique salt-stripping bristles.

The name *Tympanoctomys* refers to the enormously inflated tympanic bullae of the rodents (tympanic bullae are the bony chambers that surround the inner ear of mammals). These bullae are even more inflated than those of the kangaroo rats, which themselves have extremely large bullae. What the function of the extraordinarily large bullae could be is unknown. Do they permit the animals to hear the sounds of certain predators? Do they function in communication between the animals?

Other questions remain unanswered as well. How are the mounds constructed? Only one animal lives in each mound, but surely a single animal does not build the whole mound, for each seems too large to be constructed by a single small rodent. Are mounds inherited by offspring or appropriated by interlopers? How long do the animals live? Why do they have enormous hind feet, approaching in size those of bipedal kangaroo rats? Why are the mounds so large? To avoid being inundated by floodwaters during heavy rains, or is there a more unusual reason? I noticed that the packed sandy-salty soil of the mounds seems to magnify sounds. Could these animals use the mounds, the large hind feet, and the inflated auditory bullae to communicate by drumming on the sand, in effect converting the mounds into giant drums? Perhaps they could attract mates in this manner. There is no evidence for this, but the animals inhabit mounds whose structure, extent, and mass are difficult to explain. It takes a great deal of energy to build and maintain such enormous structures and natural selection does not like to waste energy, the currency of life. The species deserves much more attention from researchers than it has received to date, for it is one of the most fascinating mammals I have encountered in any desert.

The small mammal fauna of the southern reaches of the Monte Desert along the Patagonian coast changes gradually from what is found in the north. The yellow-nosed grass mouse *(Akodon xanthorhinus)* becomes common, its scientific name describing its yellow-orange muzzle. It inhabits the low grass and scrub desert that characterizes the southern Monte. It coexists with the small gerbil mouse *Eligmodontia typus,* but as one moves south and enters Neuquén Province another species of gerbil mouse appears, *Eligmodontia morgani.* This is one of the most beautiful gerbil mice, characterized by soft, silky fur with rich tonalities of color, a short tail, and furry feet. It lives in small hummocks of fine sand. Both species can be found together, but *E. morgani* is predominantly an animal of the southern deserts, whereas *E. typus,* while appearing in the southern deserts, is much more broadly distributed.

Other animals appear in the southernmost parts of this 1,200-mile-long desert, including a most unusual marsupial, the Patagonian opossum, *Lestodelphys halli,* which inhabits the desert scrub of the southern Monte and adjacent Patagonia. This little marsupial, which weighs only 3 ounces, is a pearl gray color and has a very short tail. It is probably carnivorous, feeding on mice and other desert vertebrates. It may co-occur with the other minute marsupial of the Monte, the mouse opossum, *Thylamys pallidior,* which extends throughout the desert. In some respects the Patagonian opossum may fill a niche similar to that of the grasshopper mouse *(Onychomys)* of the North American deserts—that of tiny, but fierce, predator.

Perhaps the most unusual desert animals are the Magellanic penguins *(Spheniscus magellanicus)* that inhabit the southern limits of the Monte.

A colony of Magellanic penguins *(Spheniscus magellanicus)* in the southern Monte Desert, Patagonia. (Photo by M. A. Mares.)

Penguins live in the desert scrub up to a thousand yards from the cold South Atlantic Ocean. The birds walk into the desert from beaches alive with South American sea lions *(Otaria byronia)* and South American fur seals *(Arctocephalus australis)* and dig burrows in which they lay eggs and raise their young. Penguin rookeries may harbor more than a million birds. Walking through a colony is an extraordinary experience, for seldom does one see a million living creatures at one time. Moreover, in such a large colony, the cycles of life and of death are evident. Everywhere one looks there are squawking and honking penguins, with young birds that are nearly fully fledged standing hungrily outside their burrows calling for their parents. There are also thousands of dead penguins—animals that were abandoned by their parents, or that died of disease or starved to death—littering the ground. The smell of fish and dead penguins is almost overwhelming , and everywhere, absolutely everywhere, are living, complaining birds. The desert soil seems to extrude penguins as they emerge from their burrows into the brilliant southern sun. It is the only place in the world where one can find penguins, camels, and desert rodents inhabiting the same patch of desert, and doing so with the roar of sea lions filling the air.

A large male South American sea lion *(Otaria byronia)* (at left, with mane) and his harem in a colony on a beach at the edge of the Monte Desert, Chubut Province, Argentina. (Photo by M. A. Mares.)

The Monte begins almost at the Tropic of Capricorn in the north, where the summer days are long and hot and the winters are short and mild, and terminates in the cold, arid wastes of southern South America, where the winter days are short and cold, and the summers short and mild. Yet across this extensive belt of aridity, through all of the isolated valleys, the small mammal fauna does not vary greatly. The mammals themselves do not perceive the different parts of this desert as being very distinctive. Indeed, there

The scrub desert of Patagonia, Chubut Province, Argentina. Patagonia strongly resembles the North American Great Basin Desert. (Photo by M. A. Mares.)

are more differences between the faunas of the northern Monte Desert and the high grasslands near Andalgalá, only 10 miles from the desert's edge, than there are between the faunas of the southern and northern limits of the Monte Desert, which are more than 1,200 miles apart.

As 1972 drew to a close, I completed my first two years of fieldwork. I had conducted a mark-recapture study on *Eligmodontia typus* in Monte Desert habitat near the city of Mendoza—the only place that I found the species to be relatively common—trapping, marking, and releasing animals on two large (about 3-acre) grids of traps for more than a year. The data gave me information on movements, habitat selection, reproduction, growth, annual cycle of abundance, dispersal of adults and young, and other ecological parameters of this previously unstudied species.

I had examined the physiology of a large number of desert and nondesert rodents in the modern physiology laboratory in the city of Mendoza that was operated by the American physiologists studying desert amphibians. I had access to a vapor pressure osmometer to measure blood and urine concentrations and other equipment that measured water balance and blood electrolyte content. When I brought animals to the laboratory, I was able to control their water and food intake, while measuring water loss and electrolyte balance during different degrees of water stress. The equipment permitted me to collect blood from the animals to determine their nitrogen and salt levels, an indication of the ability to resist aridity. I also measured water lost to excretion, urine concentration, and other things that had been found to be important adaptations for desert rodents. The data were useful in determining how physiologically adapted each species was to life in the desert.

Several other research projects had also consumed my time during these two years. I had conducted extensive fieldwork on the population ecology of tucu-tucos, learning about the burrow structure, calls, behavior, and population structure. Data from my many extended field trips had delineated the distribution and habitat selection of most of the mammal species of the northern half of the desert. Extended laboratory research had examined patterns of growth and reproduction of a half dozen species in captivity, including the growth of newborn animals. Finally, I had obtained an understanding of mammal diversity among the major habitats of northwestern Argentina, both desert and nondesert.

One of my final trips before leaving Argentina, perhaps for good, I thought, was a return to Andalgalá, where my research had begun. I traveled down the Amanao River for old time's sake, recalling my field trips in the area over the preceding two years. I set traps both along and in the dry riverbed, thinking it unlikely that I would ever return to that desert. The next morning as I picked up the traps I gave way to hubris. After working in the Monte for so long I thought that I had such a good feel for its mammals that I could predict what species would be found in any particular trap. Seeing that a trap was closed under a mesquite tree in the sandy riverbed, I knew it had to contain *Graomys griseoflavus*, the leaf-eared mouse of the mesquite woodlands, for no other small mammal frequented that habitat. The gerbil mouse did not enter the deep gullies, and no other rodents were found here. When I picked up the trap, however, I was surprised to find that the animal inside was, indeed, a gerbil mouse. How typical of fieldwork, I thought, that the last trap I was ever likely to set in the Monte Desert would have a species in it that I had never before caught in this forested riverbed. Fieldwork never ceases to surprise.

I transported the animal back to Tucumán so that I could examine it more closely in the laboratory there. Eventually, I sacrificed the mouse and began to prepare it as a museum specimen. Suddenly I realized that it was not a gerbil mouse at all. It was no animal I had ever seen. I knew at that instant that I had discovered my first new species of mammal. Moreover, I was certain not only that it was a new species, but also a new genus, for I had encountered no similar genus in Argentina. I was stunned. I was preparing an undescribed species. In the future, I would make every effort to eschew egoism. Humility is a much safer attitude with which to approach nature, for nature holds more hidden surprises than we have yet imagined.

Discovering a new species is not that rare for a biologist. New species of insects are named every day. But this was a new genus and species of mammal, and new mammals are not discovered every day. Between 1908 and 1958, for example, only forty-one new genera of mammals were named in all of the Neotropics—and the rate of discovery was in decline. I would guess that most of my friends who are field mammalogists have never seen an undescribed species outside of a museum collection. New species in North America are exceedingly rare because of the great amount of research that has been conducted on the fauna, and few U.S. mammalogists work in foreign countries. It was exciting and humbling to see new, unnamed life, and to have been fortunate enough to discover it.

My two years in Argentina had made it possible for me to collect in areas where no mammalogist had worked before, but I was an ecologist, not a taxonomist, and naming new mammals was not my specialty. I put the animal in a museum drawer and proceeded to work on the ecological and physiological data I had gathered. I left the naming of the new animal to others. It would not be until 1975 that a colleague was able to return to Argentina and collect additional specimens. In 1978 we described the new genus and species of mammal, naming the genus for the small town of Andalgalá (*Andalgalomys*, which means the mouse from Andalgalá) and the species for my friend and colleague Claes Olrog. The new mammal finally had a name: *Andalgalomys olrogi*.

When I returned to the United States, I spent four months carrying out ecological research in the Sonoran Desert of Arizona. The fieldwork was the North American portion of the comparative research that was a part of the International Biological Program. With my two years of fieldwork completed in Argentina, I now viewed the North American fauna with new eyes, for I was now familiar with a very different desert. I knew that not all deserts were alike. I also knew that deserts that look alike in one parameter, for example plant structure, could differ greatly in another parameter, such as their mammals.

My research was still designed to look for similar patterns of ecological and evolutionary adaptation, but I now had the modifying influence of experience with a different fauna on a different continent. My doctoral thesis would include the natural history and physiological research that I had conducted on the unknown fauna of the Monte Desert. But I was still interested in the larger question of how and why the fauna of the Monte evolved as it did. Was it, or was it not, similar to the fauna of the Sonoran Desert? Additionally, how similar were the patterns of diversity and adaptation between the ancient guinea pig group and the more recent South American field mice group? These questions of whether or not organisms adapted to the desert in the same way were the underlying questions of the whole IBP project, which sought to document whether or not convergent evolution occurred on a broad scale above the species level.

Two major variables were at play in the evolution of the mammal fauna of the Monte Desert: origins and time. The two major groups of rodents I was considering had different origins and probably came from ancestors that had evolved on different continents. The caviomorphs (the mara, tucu-tucos, cavies, and octodontids) began perhaps thirty million years before the sigmodontine field mice appeared. By comparison, the latter would seem not to have been exposed to deserts until they entered South America from North America a few million years ago.

There was a problem with this scenario. My hypothesis that at least some elements of a desert fauna should converge on the highly desert-adapted kangaroo rat type (a pattern reported from deserts throughout the world) did not seem to hold for the Monte. There were no kangaroo rat equivalents in that desert, either among the caviomorphs or among the field mice. If time was a major factor, the caviomorphs had had more than enough of it to develop as classic desert specialists. Other mammals had done so in other deserts. As a group, the caviomorphs were ancient, much older than the kangaroo rats or any of the "typical" bipedal desert specialists of other deserts, such as the jerboas of the Sahara Desert, which strongly resemble kangaroo rats.

I refined my understanding of the species of the Monte and their level of specialization to the desert. I surmised that the reason there were no bipedal South American field mice is that field mice had arrived on that continent too recently to begin the long process of becoming bipedal and adapting to the harsh desert habitat. The gerbil mouse, however, appeared to be moving in that direction. It looks very much like a desert specialist, with its silky fur, long tail, pale blond coloration, and long hind legs. For a time I thought that it might be bipedal, but it almost always scampers on all four feet, rather than hopping on its hind feet like a kangaroo. (I say "almost" because twenty-five years after I completed my doctoral research, while I was studying gerbil mice in Mendoza, one escaped on a barren, salt lake. As I chased the animal it made several bounds on its hind feet without using the forefeet at all. This was the first time this behavior had been observed. For those few moments *Eligmodontia* was a bipedal rodent. Thus *Eligmodontia* seems to be heading toward an evolutionary point where it could become the bipedal desert rodent of the Monte.)

To examine the morphology and ecology of the desert rodents of the

Monte and Arizona in a quantitative manner I gathered numerical data on each of the desert species. Dozens of morphological traits that reflected ecological function were measured, such as relative hind foot length, which reflects how animals move; tooth structure, which indicates food habits; dorsal and ventral color; ear length; degree of inflation of the tympanic bullae. And then the species were compared to determine their degree of similarity. But first I needed a standard to which both groups could be compared so that I could tell how similar they were to each other. So I set up a four-way comparison. I included data from desert rodents of the Sonoran Desert, as well as data on nondesert North American rodents (species from the coniferous forests of New Mexico, which I had studied over the years). These two groups are closely related, even sharing some species. Similarly, I gathered the same data for Monte Desert rodents and for nondesert rodents to which the desert rodents were closely related (I used the rodent fauna of the moist forest of Tucumán Province, which I had surveyed).

If convergent evolution had occurred at a level other than a one-to-one equivalence of species I would expect the desert faunas, despite their different origins, to be more closely allied to each other in many characteristics than either desert fauna would be to its most closely related forest fauna. Thus if desert species (regardless of whether or not they were closely related) were more similar to each other than they were to their closest nondesert relatives, this would suggest that the force of natural selection led to similar adaptations developing among animals exposed to similar environments.

When the analysis was completed, the desert faunas were closely grouped, whereas neither desert fauna was grouped with its nondesert relatives. This showed that convergent evolution had occurred across the many morphological/ecological traits that were measured for the mammals of the two deserts, but it also made clear that there were almost no exact equivalents in the two deserts. It was no surprise that the mathematical analysis closely grouped the North American pocket gopher and the South American tucu-tuco—indicating almost perfect equivalence—since the animals were remarkably alike in so many ways. Perhaps more important, when I examined grouped data of the universe of all traits measured, the analyses showed that there was strong convergence between the characteristics that had appeared among the desert species. This was especially interesting for it indicated that although one-to-one equivalents were rare, natural selection had clearly led to whole groups of similar adaptations developing within desert faunas.

My research showed that mammal faunas developing in deserts converge in many characteristics. Desert mammals develop similar body shapes, food habits, fur color, kidney physiology (the elongate renal papilla that allows them to produce concentrated urine, thus reducing water loss), adaptations of the skull and skeleton, and dental structure, as well as similarities in many other traits. But I still had not answered the question as to why no bipedal species had evolved in the Argentine desert. If the reason that the South American sigmodontine field mice did not develop the enormous number of traits that are required of a desert specialist was that they reached the desert too recently, why did the caviomorphs, with their long desert tenure, not make the leap to bipedality? Why would the Monte be the only major desert in the world to lack a bipedal species?

The answer, I thought, was that there had to have been a bipedal desert rodent in the Monte at some time in the past, but it was now extinct. If some early inhabitant of the desert had filled the bipedal desert-specialist niche, then a caviomorph would be unlikely to develop in that direction owing to competition with the animals that already filled that niche. Such reasoning agrees with competition theory, which says that animals in nature that belong to different species will avoid becoming too similar because doing so increases competition between the two species, making survival more difficult. Unfortunately, I had no proof that a bipedal mammal had ever existed in the Monte, although this did not stop me from speculating on the existence of such a phantom during my thesis defense—an argument that met with great skepticism, since it could not be proved.

In 1973 I read an article by the illustrious paleontologist George Gaylord Simpson that had been published in 1970 but that I had not unearthed in my research because I had been living in the field in Argentina for two years. The importance of the paper did not register in my mind and I did not include it in my thesis. Moreover, it was not about rodents, but about a strange group of extinct South American marsupials known as the argyrolagids. Simpson described this anomalous group at length, including pictures of the bones, drawings of the skeleton, and an artist's rendition of what the animals might have looked like in life. They had teeth that were very different from those of most marsupials. In fact, the teeth were almost exactly like those of rodents. The mammals were about the size of a large kangaroo rat. They had a long tail, reduced forelimbs, long hind legs, and large hind feet. They were definitely bipedal.

The animals had disappeared in the Pleistocene, only about a million years ago, and had inhabited the Monte Desert near Andalgalá. Here was my bipedal desert specialist of the Monte Desert, only instead of being a rodent it was a marsupial. Marsupials are abundant and diverse in South America and were present when the deserts first formed. If I had arrived in Argentina a million years ago, instead of in 1970, I would have caught them in my traps and seen at once that they were strong ecological equivalents of the classic desert rodents of the world. I had simply gotten there too late.

The several species of bipedal marsupials that had gone extinct in the southern desert probably did not specialize on seeds. They were similar to kangaroo rats and jerboas in external morphology, but their teeth were more like a jerboa's than a kangaroo rat's, and jerboas eat underground plant parts more than they eat seeds. After the argyrolagids disappeared, their niche was not filled by any other bipedal species. Did this show that convergence was predictable or did it highlight the role of chance in the evolutionary equation? I think that both chance and predictability were at play. Although these bipedal rodent-like marsupials were gone from the Argentine desert for reasons that remain unclear, at least some species of bipedal mammals had evolved there at one time.

Learning about the argyrolagids finally allowed me to show that convergent evolution had occurred at various levels across the unrelated faunas of the two widely separated deserts. The extinct argyrolagids were marsupial equivalents of the kangaroo rats. My data were also consistent with the hypothesis of a late arrival of the sigmodontine rodents from North America after the Central American land bridge had formed, the late arrival perhaps not permitting great specialization to desert life.

Some years later I was able to spend time visiting with G. G. Simpson in Tucson. Simpson had the argyrolagid fossils at his home and I was able to examine them carefully. Clearly, these marsupials were rodent-like in habits and strongly convergent on the bipedal rodents found in several deserts of the world. I held a specimen that had been collected near where I had discovered the new genus *Andalgalomys*. The fossil I was holding had existed perhaps a million years earlier in the same valley in which I had worked. For unknown reasons the argyrolagids entered the dark tunnel of extinction, while the little mouse from Andalgalá persisted in that isolated valley. As I held the fossil I felt for a moment that I could look back through time and see the argyrolagid and Olrog's desert mouse foraging together along the dry gullies of the Monte. In a sense, I bridged the eons of time between their moments of existence. I was finally able to see the bipedal "rodent" of the Monte, even though it was a fossilized marsupial that had disappeared forever from the arid lands of Argentina.

Desert in the Sky

8

They can be described only by negative possessions; without habitations, without water, without trees . . . They support only a few dwarf plants . . . Why then . . . have these arid wastes taken so firm possession of my mind?

Charles Darwin, *The Voyage of the Beagle*, 1839

It is impossible not to be aware of the massive mountains when you are traveling through the Monte Desert. In Andalgalá, for example, there is an 18,000-foot snow-covered peak, Aconquija, just northeast of town. To the west are the pre-Andean chains, and on a clear day the Andes can be seen 200 miles away. To the south is the Sierra de Velasco of La Rioja Province and to the east is the Sierra de Ambato, with peaks approaching 15,000 feet. There are imposing mountains throughout the Monte north of San Juan Province, and the desert lies cradled between them. From southern San Juan Province southward through Neuquén Province, the Monte abuts the main Andean chain, where peaks rise to heights of more than 20,000 feet. Only in the southern Monte, where the desert curves toward the Atlantic Ocean, are mountains no longer visible.

The high desert is exceedingly sparse, but scattered throughout the arid landscape of snow-clad peaks, barren rocky hillsides, and vast spare salt flats are oases that develop around water catchments. Here grow tall bunchgrasses and sedges—an emerald island within a sun-blasted world. These small areas of green may support small mammals, including rare species of grass mice (*Akodon*) and other mammals that cannot survive for long in the surrounding barren landscape. My colleagues and I have even discovered new species in these areas. The oases, whether of fresh water or salt water, provide moisture and breeding grounds for large mammals such as the guanaco *(Lama guanicoe)* and the delicate vicuña *(Vicugna vicugna)*, Andean camels that inhabit the highest deserts on the continent. The vicuña is prized for its wool, which is among the finest in the world. Both species were hunted by the Incas and later driven nearly to extinction by modern hunters in search of wool for native weavings. Recent efforts at conservation have brought the animals back from the brink and they are again common in portions of the high desert. In the Laguna Blanca reserve (white lagoon, so named because it is a salt lake) in Catamarca Province, vicuña populations have been estimated at more than 20,000, and we saw thousands each day running free in the vast protected area.

A lone vicuña at 13,000 feet above sea level in the high desert of Catamarca Province, Argentina. (Photo by M. A. Mares.)

The mountains of Argentina are not like the Rockies, for they support little vegetation and almost no trees. Only on the eastern faces of the eastern mountains of northwestern Argentina—the sides facing the great thorn forest—do the mountain slopes support dense, green forest. In Tucumán the lush subtropical forest of the lower slopes and the alder forest of the higher elevations make the mountains appear to be wearing a thick, green robe. The western slopes of these same mountains are arid, their dry hillsides supporting only a few shrubs, cacti, and ground bromeliads.

The mountains north of Andalgalá provide a good example of the high-mountain vegetation bordering much of the Monte. The town of Santa María, Catamarca Province, lies only 70 miles north of Andalgalá, if the distance is measured on a map. But in the Northwest map distances are deceptive because rugged mountains lie in all directions. There is a road leading from Andalgalá to Santa María that has been carved from the rocky face of the mountains that separate the two desert towns.

The little-used road provides access to a rhodachrosite mine. It ascends via the Cuesta de Capillitas, "the mountain road of the small chapels." The road, along which there are no chapels, is a marvel of engineering, especially impressive because it was built at the turn of the nineteenth century. It is a narrow dirt track that climbs continuously upward, utilizing hundreds of switchbacks, to an elevation of 10,000 feet. In places it is carved from rock with sheer drops of more than 1,000 feet. The road begins climbing amidst a dense forest of cardón cacti, the saguaro-like sentinels of the desert. Eventually, it rises above the limits of cacti and the habitat changes to small shrubs and bromeliads. Finally, thick bunchgrasses predominate.

Andean condors (*Vultur gryphus*) appear—enormous birds with a wing span of 10 feet and a fluffy white collar setting off the jet-black body. They have a striking white patch above each wing and as they twist and turn high in the sky their color shifts from stark black, to startling black and white, then back again to black. Once a group of more than a dozen birds soared around me as I stood on a promontory at the edge of a precipice 1,000 feet high. The condors were gliding just above my head and I could hear the loud whoosh of the wind rushing through their stiff primary feathers. Condors seldom beat their wings and expend little energy in flight. They make delicate ad-

The mountain road (view to the south) leading from the Monte Desert to the montane bunchgrass prairie and high desert near Andalgalá, Catamarca Province, Argentina. The Salar de Pipanaco, a great salt flat, is visible in the upper right of the photo as a white area extending southward into the valley. (Photo by M. A. Mares.)

justments to their outer wing feathers as they control the winds of the desolate mountain canyons. They can remain aloft for hours before returning to earth after lonely flights of hundreds of miles over the barren Andean landscape.

There is a tiny settlement with a small school above the Capillitas mine in the high scrub and grass prairie at 11,000 feet above sea level. In the early 1970s the arrival of any vehicle in the area, in this case one containing my students and me, was so unusual that class was suspended. The teacher and the students—who were from eight to twelve years old—gathered at the front door to peer out at the strangers. I walked up to the young teacher and told her that we were biologists who were studying the mammals of the area. She invited us into the classroom and asked us to share with the class the purpose of our trip.

It was a small room, but the building was solidly constructed, clean, and neat. There were few amenities. The school had no electricity, but a small wood stove provided heat. The teacher, who was the only teacher at the school and taught all the classes, had few teaching aids. A poster near the front of the room caught my eye. I struck up a conversation with a young boy about what was shown on the poster, including bats and other mammals and a variety of trees.

The bat pictured on the poster was an Old World bat, rather than one of the sixty or so Argentine species. I asked him if he had ever seen a bat. He replied that he never had. I did not find his ignorance about bats surprising, for few people ever see those shy, nocturnal denizens of caves and abandoned buildings. Moreover, bats do not do well in the cold, high Andes. Only one or two species would be expected to occur in the area, and these would be rare. Without thinking much about it, I asked him what another thing depicted on the poster was, and I pointed to a tree.

"It is a tree," he said.

"Right," I said, "Have you ever seen a tree?"

"No," he replied, "but my father has seen them. He says they are very beautiful."

The isolation of the small settlement, where the howling winds, cold winters, aridity, and high altitude made it impossible for trees to grow was driven home to me. Not to have ever seen a tree! I had never before met anyone who had never seen a tree. I felt like presenting the boy with a translation of Joyce Kilmer's famous poem, for here was someone who really would appreciate, at a most fundamental level, the singular beauty of a tree. Like Kilmer, but without his uncertainty, this child would have known that, indeed, no poem could be as lovely as a tree.

The view through the window was of the treeless Andes, their snowcapped slopes visible in the distance, the windswept grassy valleys surrounding the school reaching to infinity. How sterile the tree appeared on the poster, a two-dimensional representation of a three-dimensional, complex living or-

Windswept bunchgrass prairie in the high mountains above Andalgalá, Catamarca Province, Argentina. (Photo by M. A. Mares.)

ganism. It appeared even more lifeless than the bat. What would this young boy say about the great Amazon forest or the hardwood forests of the eastern United States? His lack of familiarity with the natural world made me feel privileged to have experienced so much of life's diversity.

"Some day I will travel to the edge of the grassland and see the trees," he said, "for my father says that they grow there. Some day I may go to Tucumán."

Tucumán, the "Garden of the Republic." There were certainly trees there—a viridescent forest that cloaked the mountains west of the city. A folk song has described the rich forests of Tucumán as being painted with a thousand shades of green. It is a complex forest that harbors parrots, toucans, liana vines, bromeliads with flaming red flowers, and a plethora of hummingbirds and other species that depend on the forest for life. These species were notably absent from the altiplano (high plain), the high desert in which we found ourselves, a desert island in the sky that is separated from the lowlands by the miles-high Andean slopes.

We had been in a desert woodland near Andalgalá only two hours before, and if we continued on to Tucumán we would be in the pale green alder forests within another three hours, and in the complex wet forest an hour after that. But for someone who had to travel on foot the great forests of Tucumán demanded a trip as long as that undertaken by the Apollo astronauts when they journeyed to the moon.

I had no doubt that the boy would reach the forest some day, for he had a goal. His vision already extended well beyond the tough bunchgrass surrounding his home. His father had lifted his dreams to a further horizon, as fathers are supposed to do. His journey would be no less challenging than that of the astronauts. The poor walk everywhere in Argentina, and the boy would have to walk to the forest, a trip that might take a week. In some ways he was like a child in the United States dreaming of going to the moon, for the forests of Tucumán were no less remote to him than the moon is to a North American child.

The remote altiplano of the Andes is one of the most uninhabited places on earth. There are seven people per 100 square miles in this part of Catamarca Province, less than one tenth the population density of Alaska. It is a hard place to eke out an existence and only a few poor hamlets dot the altiplano. The settlements are inhabited by a mix of Indian and European peoples, but most are Indians. Some are descendants of the Incas and many speak Quichua, the language of the altiplano from Argentina to Ecuador. Argentines call them *collas*, natives of the high country. Their music, played on *bombos* (large drums), *charangos* (the small "guitars" made from the shell of the screaming armadillo), *quenas* (bamboo flutes), and *zancas* (wind instruments made of different lengths of bamboo, each with a different tone, tied together into a bundle), hauntingly reflects their isolated lives.

In March of 1982, when Argentina invaded the Falkland Islands, which the Argentines claim were stolen more than 150 years ago, they entered into a war with Great Britain. It was an unwinnable war for Argentina, but the losses on both sides would have been greater if Argentina had actually fought for the islands with its best troops and supplied them with proper equipment and sufficient ammunition. The way in which Argentina conducted the war was shameful, and only the fact that it lasted a short time kept it from being a bigger farce than it was. Eventually, the corrupt military regime in Buenos Aires collapsed, and democracy—that small plant with deep roots that is both delicate and tenacious—returned to the country, first tenuously, then with greater spirit.

When it came time to send troops to die for these cold, windswept, tree-less, and sea-washed islands, the Argentine military opted to send new train-ees who were considered expendable. Among the young boys chosen were collas from the altiplano of the Northwest, Argentina's forgotten people. The boys were given perfunctory training and sent on ships to do battle in the Falklands. There they would fight the highly trained British military, includ-ing the famed Gurkha brigade.

The only way Argentina could transport troops to the battle zone was by ship, for the British Navy had established a defensive antiaircraft ring around the islands that made it impossible for Argentina to fly in troops. The Argen-tine Navy was hopelessly out of date. Its main troop ship, the *General Belgrano*, was an aging hulk that had once belonged to the United States (when it bore the name *U.S.S. Phoenix*). It had even survived the Japanese attack on Pearl Harbor. Eventually, Argentina purchased it and converted it to a troop transport. This slow, defenseless ship was chosen to transport the soldiers from the Argentine mainland to the desolate islands engulfed in war.

On May 2, 1982, in the depths of the Argentine winter, the ancient ship was sunk by the British submarine *Conqueror*. That day 368 young men were lost in the frigid waters of the South Atlantic, many horribly burned in the flaming oil from the wreckage. Some were boys from the altiplano. They had never before seen the ocean. Some had never seen a tree.

The Vampire and the Phantoms of All Hallows' Eve

The earliest attempts to explain the natural world invoked
the supernatural. From the most primitive animism to the
great monotheistic religions, anything that was puzzling
and seemingly inexplicable was attributed to the activities
of spirits or gods . . . But what really is an explanation?
When a puzzling phenomenon is encountered in the ev-
eryday world, most frequently it is "explained" in terms of
what is known or what is rational . . . But merely having a
rational explanation is not enough.

Ernst Mayr, *This Is Biology*, 1997

When the Nazca Plate, the massive geological
formation upon which much of the eastern Pacific Ocean rides, collided
with the American Plate, which carries South America, the enormous forces
released heaved up the mighty Andes Mountains. This series of titanic
mountain ranges extends unbroken from Venezuela in the north to Tierra
del Fuego in the south, a distance of more than 5,000 miles. The mountains
in the chain dwarf the mountains of North America and Europe. The land
along the western parts of South America was raised miles into the air,
and many peaks reach almost 4 miles in elevation. Mountains with peaks
of 14,000 feet, which would be awe-inspiring massifs in North America,
are hardly noticed in South America, especially in northwestern Argentina.
Many Andean desert valleys lie above this elevation.

In western Argentina the mountain peaks are snow-capped all year round and cloaked entirely in deep snow in the wintertime. Most mountains in the Andes are oriented from north to south, but in some areas mountain chains aligned on an east-west axis meet the north-south chains and form isolated valleys. The town of Andalgalá is in such a valley, at the base of a mountain 18,000 feet high called the Nevado de Aconquija.

On October 31, 1972, Halloween in the United States, I decided to spend the night collecting bats over the narrow Potrero River, about 8 miles north of Andalgalá. My field crew set out in the late afternoon for a spot where I had previously been successful in capturing bats, a place where a small dam formed a tranquil waterhole. The bats drank on the wing as they flew down the river toward town, where they would feed on the insects attracted by the crops painstakingly irrigated and cultivated in the lowland desert soil.

The river, hardly more than a stream, flowed between steep hillsides covered with thorn scrub, large cacti, and stunted desert trees. From the point where my crew and I parked the car, we had to walk several hundred yards to reach the dam. The deep azure sky of late afternoon became crystalline and the shadows of the hills lengthened, darkening our path as we hurried to arrive in time to set up the mist nets before the first bats began to fly. Only the gurgling of the water broke the stillness of the warm desert air.

As we wended our way up the path toward the dam I chanced to look up at the sky. I stopped suddenly and stared at three massive globes silently floating just above the dark ridge to my left, one red, one green, and one blue. I called everyone's attention to this inexplicable sight. Like the Wise Men gazing at the Christmas star, we were reduced to murmuring words of wonder.

There we were, five biologists trained in science, with years of postgraduate education between us, heads thrown back, jaws agape, rooted to the dark trail in the small, isolated valley, each of us trying to identify something we had never seen before. We began to exclaim, "What are they?" "How high are they?"

One of us said, "They look like they are only a few feet above the trees on the ridge."

"No, they seem to be higher than that, perhaps a few hundred feet above the ridge."

I remembered that I had left my binoculars in the car. I said I would go back to get them.

"Alone?"

"Well, yes."

One person said, "I'd better go with you."

"Sure," I said, "that way they can beam both of us up at the same time."

By now darkness had fallen. "There's got to be a logical explanation," I said as we started down the trail to the car, but I could think of none. Our edginess and discomfort at this confrontation with the unknown increased as we neared the car, which appeared to be directly under the largest and apparently closest globe, the red one.

Having encountered the inexplicable, we floundered around trying to fit the unfamiliar into a scheme of the familiar, trying to adapt our worldview to include the bizarre. During World War II the forest people of New Guinea concluded that the warplanes they saw overhead were giant birds, placing the inexplicable in a framework of knowledge and experience that they understood. We were in much the same position except that we were not primitive islanders. We were biologists with strong backgrounds in science, working in an isolated part of the world. We had only to reason this occurrence through in order to understand it. That, of course, did not keep the hair from rising on the back of our necks as we walked the lonely trail back to the car.

I retrieved my binoculars and looked up at the globes. Even with magnification I could not tell whether they were hovering high in the sky or floating just above the trees. They could have been a hundred feet above the ridge or thousands of feet above. It was impossible to tell. We returned to the dam, where the others were placing the fine mist nets across the water to ensnare the bats.

We went to work, but often paused to look up at the bright orbs sitting still and silent overhead. Then gradually, as we watched, they moved slightly. Then, almost miraculously, the green one moved into the blue one, so there were now only two. Then the blue one moved into the red one, and only one globe remained. It continued to glow and even seemed to expand; and then its glow began to fade. We watched transfixed as it disappeared.

We had witnessed a remarkable event, yet we had no idea what we had seen, how it occurred, why it began, or why it ended. We did not know how large the objects were, how high they were, or how they moved, since we could hear no sounds that would indicate a power source.

So eerie was the situation that someone said apprehensively, "You know, midnight is the start of the Day of the Dead." True enough. The first of November is All Souls' Day, when Catholics, particularly those in Latin America—who are extraordinarily interested in all aspects of death—are one with their dead ancestors. In the isolated Andean villages, that is a time of ceremony and mystery. But surely, I thought, there has to be a natural, not a supernatural, explanation for what we had seen.

"Maybe," I said in a weak attempt to restore a sense of reality, "there is something in the atmosphere—maybe like water droplets make a rainbow."

"Those colors didn't come in the order that one finds in a rainbow, you know. When did you ever see a rainbow like that?"

Point well taken. No rainbow would appear as three separate and well-defined colored globes.

Something must be reflecting light, but what? None of us could proffer a hypothesis that made any sense. Part of the problem was that we had no idea how high the globes were. They seemed low enough that we could climb up the mountain and perhaps get close enough to find out what they were. But the mountainside was steep and covered with cacti and spiny bromeliads. It would have been an extremely difficult climb even during the day. At night it was not possible.

"Maybe others saw this too. Maybe they were high enough in the sky that people in Andalgalá saw them. Then again, maybe we'll never know what they were."

The wine that we often carried with us for the all-night session of bat netting flowed freely. We speculated about death, flying saucers, and all sorts of strange phenomena. Someone mentioned spontaneous combustion, and we were off on long discussions of every puzzling happening we had ever read about, as adults or as children. Throughout that long night we somehow managed to catch a few bats, including a vampire, only the second one ever found in the valley. A rare vampire on the Day of the Dead seemed appropriate.

We returned to Andalgalá the next morning, weary after the exertions and strange experiences of the night. We learned that many people in the small mountain villages had come screaming with fear into the streets as the globes suddenly appeared above their towns. Some had fallen to their knees filled with terror, rosaries in hand, making the sign of the cross. The coinciding of the glowing orbs in the dark sky with the Day of the Dead was not missed by the religious and superstitious. Some undoubtedly thought the end of the world was nigh. Many surely repented their sins and prepared to meet God. But no one could offer a rational explanation of what we had seen.

A few days later I returned to Tucumán, a modern city of several hundred thousand people situated on the east side of the mountains, with nothing between it and the capital of Buenos Aires but flat thorn scrub and grasslands. In Tucumán, unlike Andalgalá, there were newspapers, television, and radio. I bought a newspaper and read with great interest a small news item on a back page concerning atmospheric testing done by the Argentine space agency. The agency had wanted to test the properties of the highest levels of the atmosphere by blasting a rocket into space. The rocket contained three colored gases—red, green, and blue—and they were released from 120 to 200 miles up in space, forming spheres more than 100 miles in diameter that were visible for more than 1,000 miles.

It was remarked that the experiment had caused concern among some residents of the countryside, particularly the superstitious people of small, mountain villages, who were not trained in science and certainly were not aware of the origin of the bright balls of gas. Because of their great height, the glowing globes reflected the rays of the setting sun well after dark. The space agency was well satisfied with its experiment.

Like the unschooled people of the mountain hamlets, we did not even know that Argentina *had* rockets, much less rockets that could reach such a tremendous altitude and then expel enormous balls of colored gas. To us the gas balls appeared to hover just above the trees, within range to collect a few human specimens as we were collecting bats, perhaps with no more thought given to our unique natures than we paid to the unique nature of the bats. The bats and we were one before the unknown. And, in retrospect, our ignorance was more frightening than the eerie spheres that glowed above the dark valleys of the Andes on the eve of the Day of the Dead.

Land of the Shah 10

There are a few animals whose adaptation to deserts sur-
passes that of both the plants and amphibians. These are
the kangaroo rats, pocket mice, and jerboas. These desert
rodents have developed similar characteristics in different
parts of the world, but they are not all the same species and
even belong to different families . . . Desert animals . . .
never dominate the landscape. The traveler who honestly
can report many species must be either a most alert ob-
server or a very lucky person, for each creature sighted and
identified marks a red-letter day of a desert journey.

Alonzo Pond, *The Desert World*, 1962

I was able to tie the story of the comparative
ecology and evolution of desert mammals in the Argentine Monte and the
Sonoran Desert together into a neat package, but I had nagging doubts
about the results. Clearly, the two deserts were much alike in climate, plants,
and geology, but the mammals were less similar than I had anticipated. I
now knew about the extinct kangaroo-rat–like marsupials, so convergent evolution seemed to be both significant and repeatable as mammals evolved in
deserts. Additionally, my mathematical analyses of morphological traits that
reflected ecological function showed that when the characteristics of the
mammals of each desert were considered together, convergent evolution
had occurred on a broad scale: similar traits had evolved among many species in both faunas. The only odd finding was that very few "perfect equivalents" had evolved, those species that for all intents and purposes are re-

flections of each other, even though they are unrelated and had developed in different deserts.

Time appeared to be a key factor. The guinea pigs and their relatives, which entered South America tens of millions of years ago, had not developed kangaroo-rat–like species because the marsupials had already filled the very specialized niche the rats might otherwise have filled. The field mice that entered from North America seemed to have arrived in South America too recently to have developed the great specializations that are required of a species that intends to be as highly adapted to the desert as the kangaroo rat. How could I tease time out of the equation? Perhaps by going to a desert where the mammal fauna had been associated with the desert for as long a period of time as had the North American kangaroo rats with the North American deserts, I would be able to test for time as the critical variable. I searched for a desert that would permit me to find an answer.

In 1975 I went to Iran. The Iranian Desert is part of a complex band of deserts that cuts a wide swath across Asia from India to the Middle East. Iran is mostly a desert country, and the northwestern desert is called the Dasht-i-Kavir (*dasht* means desert and *kavir* refers to the vast salt flat that lies within that particular desert). The Dasht-i-Kavir is a desert where few trees grow and looks like the driest parts of the Sonoran or Chihuahuan deserts. Like the Sonoran and Monte deserts, the Kavir is a basin-and-range desert at about 30° north latitude, with low-lying basins scattered between isolated mountains. At first glance some shrubs resembled creosote bush, but were in the genus *Tamarix*, the same genus as salt cedar, that desert invader that has damaged large tracts of the American Southwest. Most shrubs were less than 2 feet tall, although the occasional wild almond reached 6 feet in height. Like the high arid plateau of Argentina, the Kavir was mainly a desert without trees.

I arrived in Iran as a guest of the Iranian Wildlife Service. The service had become familiar with my desert research and wanted me to assess whether a long-term study of small mammal ecology was possible in Iran. I was happy to comply. Unfortunately, I found that the bureaucratic obstacles in Iran exceeded those I had encountered in South America, and for a time it appeared that I would not get to the desert at all.

First, I was told there were no vehicles. Then I was informed there might

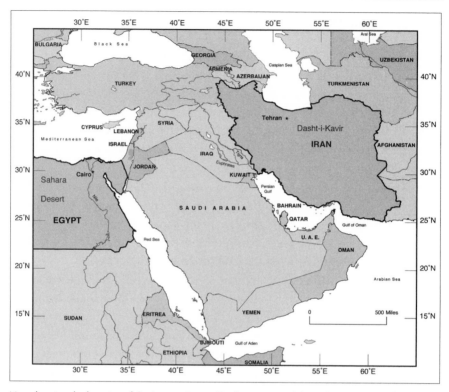

Map showing the location of the Iranian Desert (Dasht-i-Kavir) and the Sahara Desert in Egypt. (Map by Cartesia MapArt, modified by Patrick Fisher.)

be one available, but at the last moment, alas, it would not be available after all. The bureaucrats were always well mannered and tea was served at all meetings. The Iranian way of consuming tea is to take a sugar cube, dip it in the tea, and then eat the sugar cube. In this manner, the bureaucrats disposed of enormous numbers of sugar cubes and eventually finished the tea. I drank my tea without sugar, arousing great consternation. At one session more than a dozen people were called in to observe my strange behavior.

The head of the wildlife service was related to the previous Shah of Iran. He was called "His Excellency" and everyone groveled around him. He had been educated at Harvard and spoke English much more beautifully than I did. I had been unable to obtain collecting permits or a vehicle after spending more than a week in Iran and was becoming increasingly frustrated and vocal about my situation. I was in no mood for sycophancy. A large meeting was called to deal with the "problem" of the unhappy American. I was appalled that even North Americans treated His Excellency obsequiously. Eventually I said something like "Look Excellency, if you were really in charge of this place you would come up with a vehicle for my research. I do not have time to sit around listening to a string of excuses as to why I cannot get permits, or a car, or go to the desert. You invited me, so get things moving."

My comments shocked everyone at the meeting, and for a while everybody was very quiet. I was not feeling good, because I was exhausted from jet lag, lack of sleep, and raging allergies, and I simply had lost my patience. Eventually, His Excellency apparently decided to ascribe my audacity to my being an impatient and uncouth American with little breeding or understanding of the niceties of international boot licking. He said a vehicle would be "no problem," using the very words that over the years had come to be an auditory red flag to me. I began to get a sinking feeling that no car would be forthcoming. His Excellency abruptly ended the meeting. As I expected, days passed but no vehicle appeared. Then I was sent word that it might be weeks before a vehicle could be found. Tehran was becoming oppressive.

Then I learned that a biology class from Damavand College, a private school with English-speaking instructors, had chartered a bus for a one-day field trip to the Kavir protected region, the very place that I was trying to get to. A North American taught the class and I was invited to go along. I requested that they drop me off in the heart of the desert at a caravanserai, an inn of the type used by the caravans that traversed the old Road of Silk bringing spices and other precious goods from India to the Middle East. I would worry about how I would get back to Tehran later. Those in charge of the trip told me that all it would take to put my plan into effect was a letter from His Excellency. After making inquiries I was told that the letter instructing the guards at the caravanserai to give me lodging and provide a vehicle for my use would arrive by early afternoon. It would be no problem.

Of course, the letter did not arrive. The next morning when the bus pulled up it was not the twenty-seat vehicle that had been ordered, but one with only thirteen seats. It looked as if I would not be able to go after all. There was not even enough room for all the students in the class. Finally, one of the students, a thirty-eight-year old American divorcée living in Tehran, decided to take her Jeep. I managed to hitch a ride with her and two young Iranian women, all of them students in the biology class. It was an eight-hour round trip to the desert, but at least I would be able to see the area.

We drove through parts of the Kavir Protected Region. It was a beautiful desert reminiscent of both the American Southwest and the Monte. The reserve was partially fenced and great efforts had been made to keep camels and other livestock out of the park. It was probably as fine an example of largely undisturbed desert as one could find in western Asia. The vegetation consisted of low shrubs and recalled the most arid parts of the American deserts. I saw the tracks and burrows of rodents everywhere. It seemed to be an ideal place to work. We also stopped at the caravanserai, where we had lunch before beginning the long ride back to Tehran.

On the return trip the Jeep broke down as we were traveling along the main highway. I thought it particularly fortunate that the car died within sight of a service station, but I was riding with two Iranian women and a North American woman, all dressed in Western garb. I examined the Jeep, but with my minuscule knowledge of vehicles could do nothing. I thought I knew what was wrong but I did not know how to fix it. I explained the problem to the women, who all spoke Farsi. I asked them to go to the service station and ask for help. They said that if they did that they could be assaulted or even stoned, since they were not wearing traditional Muslim clothing. They tried to teach me some phrases of Farsi that would explain what was wrong with the car, but it was hopeless. I could barely explain car problems in English. I offered to walk with them to the station but they adamantly refused. Moreover, they did not want the mechanics to come to the Jeep because they thought they would be harassed because of the way they were dressed and for being with Westerners. "This is not Tehran," they pointed out.

We sat beside the busy highway for more than an hour, waiting for something to happen. Finally a military Jeep pulled up behind us to find out what we were doing in the area. The women explained to the soldiers that the car had broken down and asked if we could get a lift to Tehran. There they could find someone to retrieve the vehicle. The soldiers seemed very suspicious, but finally loaded us into their Jeep. That was one of the most frightening rides I have ever taken, surpassed in recklessness only by a cab ride in Calcutta a decade later.

Traffic in Tehran was unbelievable. The soldiers, being heavily armed and in an official vehicle, would yield to no one as they raced at breakneck speed through dense traffic, once even driving on a sidewalk to get around stalled cars, forging ahead as if the sidewalk were the autobahn. At the time, killing a pedestrian carried a mandatory prison term, regardless of fault, unless the driver was in the military. Like James Bond, the soldiers had a license to kill. I saw a civilian vehicle in heavy traffic run a red light while going the wrong way down a one-way street—backwards. During this hair-raising ride I was rethinking the wisdom of purchasing a Citroen, called a Giand in Iran, to deal with the Iranian roads and traffic. A tank would have been more useful, except for the penalty for killing a pedestrian, which was a worry. I wrote in my notes: "A Giand in this traffic seems like a ticket to disaster."

A few days later a Land Rover was finally made available for a trip to the desert. I shopped for groceries and went to the bazaar to buy plastic bags and other necessities. There I ran across a copy of a *New Mexico Magazine* many years old. It had a picture taken at Christmastime of Old Town, the neighborhood of my youth, alight with glowing *luminarias*—paper bags with candles in them that light the way for Joseph and Mary. The tranquil scene I knew so well clashed with the crowds and the cacophony of people shouting and dickering in Farsi. I was probably the first person from Old Town to see the magazine since the old fellow at the newsstand had received it years earlier. I left it on the stand, a peaceful reminder of my desert home for visitors to the noisy Tehran bazaar. After shopping, I waited for the truck and driver to arrive. I was armed with five phrases in Farsi: "turn right," "turn left," "go straight," "stop," and—just for good measure—"I do not speak Farsi." I was ready to do research in the Iranian Desert.

The caravanserai is a 300-year-old fort in the desert situated at the base of the Black Mountains. It has many rooms on two levels built around a courtyard. Camel caravans were once led into the courtyard, where the travelers could be protected from roving bandits. When I was there it was a tourist stop for the very few hardy visitors who wanted to spend a night in the desert. The rooms were sparsely furnished and very dusty. I began to sneeze uncontrollably, so I dragged my cot into the courtyard and slept under the desert sky. This surprised the guards, who much preferred sleeping in the rooms. Why would anyone want to sleep outside?

The Iranian Desert (Dasht-i-Kavir), with the caravansarai and the Black Mountains in the distance. (Photo by M. A. Mares.)

Near the caravanserai is a haremserai almost as large as the caravanserai, but with more and smaller rooms. Over the centuries, the haremserai had fallen into ruin, but in the glory days of camel caravans it was the place where the harems were lodged. They were guarded by eunuchs, who prevented anyone staying in the caravanserai from sneaking into the haremserai for a night of forbidden delights.

I awoke at 4:30 A.M. on my first morning in the Dasht-i-Kavir after having slept under a cold starry sky. I had set only a few traps the evening before in the dark, and I was not expecting many animals. As I walked through the desert I saw many signs of rodent activity—burrows and mounds, diggings under bushes, and tracks on sand. If small mammal activity was any indication, this desert was far different from the Argentine Monte and more like the North American deserts. I caught only one animal, a small Sundevall's jird *(Meriones crassus)* that looked like a kangaroo rat in the trap. It is not a bipedal species, however, though it is clearly a desert specialist. Jirds have long been known to be able to exist without free water, and, like kangaroo rats, they eat seeds. I found that another species of jird, the Libyan jird *(Meriones libycus)*, was also present. This rodent is about the size of a North American woodrat or the brush-tailed vizcacha rat of Argentina. Unlike those animals, however, this jird specializes on seeds and lives in elaborate mounds that look very much like those of kangaroo rats. It is shaped like an animal that was designed to burrow, being muscular and rather tubular in form, and having small ears and smooth fur.

The desert was very arid and there were few types of habitat. Mostly it consisted of low shrubs spaced far apart. As in the cooler parts of the North American deserts, sagebrush *(Artemisia)* was common, making the desert seem familiar. In some areas pebbles coated the surface—a formation known as desert pavement—and in others the soil was sandy. The mountains were even more arid than the lowlands, with only widely dispersed shrubs growing on their black, rocky slopes. I found an occasional spring lined by reeds and tall grasses, but these were tiny oases of green in a vast expanse of aridity.

A great salt flat that appeared to be almost devoid of life was a short drive from the caravanserai. It seemed to stretch to infinity. I found no sign of mammal activity in that area. On the way to the salt flat I saw mountain gazelles *(Gazella gazella)* racing through the scrub, reminding me of North American pronghorns. The gazelles were more delicately built, however, and had an odd habit of running a few steps, then hopping a few, then running again, a behavior called stotting.

I explored the desert surrounding the caravanserai over the next two weeks, collecting mammals, scouting out study sites, and bird watching.

Field biologists often encounter things that take their breath away the first time they see them. Once while working high in the Andes of tropical Venezuela along a roaring mountain stream, I had been marveling at the force of the water and thinking that it would be impossible for a person to cross the raging torrent—surely anyone attempting to cross would be swept onto the boulders below and battered to death. No sooner had I had this thought than a bird appeared just a few feet away from me in the midst of the deafening cascade. Paying no attention to the stream, it turned and looked at me, then disappeared under the water. The whole incident lasted less than 5 seconds. I stared at the stream, wondering whether my eyes had deceived me, and then suddenly the bird appeared again about 15 feet *upstream*. It stood for a moment on a submerged rock with tens of thousands of gallons of water roaring by every minute, then dipped under the water again. I had just seen my first torrent duck *(Merganetta armata)*, a small duck that inhabits streams in the Andes and feeds in the raging water. It was an astounding and unforgettable sight. Now, in the Iranian desert, I encountered a species that was equally remarkable.

Walking through the desert I suddenly heard a loud whistling sound, as if a small projectile had sailed over my head. I looked up but saw nothing. Soon I heard another whistling whoosh, but this time I saw that the source of the sound was a bird, one that seemed to fly like a bullet and then disappear. I watched to see where it landed, but lost sight of it as soon as it did. It seemed to be invisible on the ground. Soon a third blew past me, the air loudly whistling through its wings. This time I managed to keep my eye on it after it landed. It was about the size of a large pigeon, and its mottled brown color made it very difficult to see against the desert soil. It was a black-bellied sandgrouse *(Pterocles orientalis)*, a remarkable bird that lives in some of the most arid deserts of the Old World. Among its many adaptations for desert life, the feathers on the belly of the male are specialized to hold water. The birds are able to live in waterless zones because the males can fly to distant waterholes, fill their feathers with water, and bring the life-giving liquid back to the female and her young in the nest. I felt privileged to be seeing this unusual and highly specialized desert bird.

About a week after beginning my work at the caravanserai, I returned one morning after checking my traps and found that a student from Oxford had arrived to study gazelles. Remarkably, he was a friend of one of the people who years before had collected bats with me beneath the glowing balls of gas. I was almost out of food, but he was collecting gazelles for the Iranian Wildlife Service, studying their food habits, parasites, and internal anatomy. So we feasted on gazelle kabobs and rice that evening, a welcome relief from the flat bread covered with a thick, green mold and canned eggplant that had been my food for several days.

Small mammals were fairly abundant in the Iranian Desert, and each morning's catch seemed to yield something different. One morning I checked my traps and found yet another species of jird, the Persian jird (*Meriones persicus*), this one an animal that looks remarkably like the brush-tailed vizcacha rat of Argentina. Like the South American species it inhabits rocks and even has a tail that bears a flag of stiff hairs. This jird, too, feeds on seeds. Later that morning I encountered my first gerbil, the Baluchistan gerbil (*Gerbillus nanus*), which looks amazingly like the gerbil mouse of the Monte. This gerbil is a small mouse, pale blond above and snow white below, with long hind feet and a long tail. Like North American pocket mice, and unlike the Argentine gerbil mouse (which eats seeds, insects, cacti, and other foods), the Iranian gerbil is a seed specialist.

On walks through the desert, I saw a variety of birds, extremely different from those inhabiting the deserts of the New World. There were brown-necked ravens, chiffchaffs, white wagtails, cream-colored coarsers, red-backed shrikes, hooded wheateaters, and see-see partridges. One day I watched a crested lark hover in the air for more than a minute while singing the most complex and exquisite bird song I had ever heard. As he sang he fluttered higher and higher until at last he ceased singing and plummeted to earth, seemingly exhausted from performing his aria.

One morning I checked my traps and found a spectacular rodent, the small five-toed jerboa (*Allactaga elater*). As I walked toward the trap, the animal looked exactly like a kangaroo rat. There were the long hind legs, the enlarged hind feet, the same pale color, and the tiny forearms. But there was at least one difference: the jerboa did not have the inflated bony ear chambers (tympanic bullae) that characterize the kangaroo rat, but instead had very long, almost rabbit-like, ears. In other respects, however, it appeared to be a close analogue of the kangaroo rat.

A five-toed jerboa *(Allactaga elater)*. (Photo by K. Rogovin; Mammal Slide Library, American Society of Mammalogists.)

This jerboa belongs to the family Dipodidae, which contains many genera and species of bipedal rodents that occur across the deserts of Eurasia. It is not a close relative of kangaroo rats—the two groups have had separate evolutionary histories for tens of millions of years. Surely here was my convergent equivalent of the North American kangaroo rat, as many published reports had suggested. My research would eventually show that things are not always as they seem, though, for the jerboa and the kangaroo rat, although presumed for years to be ecological equivalents, are in fact quite different in fundamental ways.

One morning while I was checking my traps, an Iranian Air Force F-4 Phantom jet roared a few hundred feet overhead. I did not hear it coming and the sudden explosion of sound startled me. F-4's are exceptionally noisy and smoky, and for a moment the desert hills resounded with the roar of its afterburners. It was gone in a heartbeat and then silence returned. But then suddenly I heard something running toward me. Two Persian wild goats *(Capra aegagrus)*, very large goats called ibexes with long, scimitar-shaped horns, had been frightened by the plane and were running in panic down the rocky hill in my direction. I crouched behind a bush and they came to within 10 feet of me before they saw me. Apparently startled, they leaped high into the air, then bounded back into the hills. Like the desert bighorns of North America, ibexes inhabit extremely arid habitats, where they subsist on sparse vegetation. They are marvels of desert adaptation, thriving where few large mammals can exist.

On a rocky, boulder-strewn hillside, I found that I had collected several long-tailed hamsters *(Calomyscus bailwardi)*. They look somewhat like the canyon mouse *(Peromyscus crinitus)* of the Southwestern United States and weigh only an ounce. One difference between the Iranian and North American species is that the tail of the long-tailed hamster has a flag on it formed by stiff hairs, making it distinct from most other small desert rodents. Long-tailed hamsters reportedly eat seeds, although I found that their teeth were structured much like those of rodents that eat insects, including the canyon mouse.

In the evening I drove through the desert to see if I could find any other bipedal rodents. I managed to catch the greater three-toed jerboa, *Jaculus blandfordi*, a species even more like a kangaroo rat than the five-toed jerboa I had caught earlier. At first glance, *Jaculus* was hard to distinguish from a kangaroo rat. They both hopped on their hind legs, they had similar body proportions and coloration, and they had long been thought to be exact ecological counterparts. I had chased kangaroo rats through the desert since I was an undergraduate, and in Iran I finally got to chase another bipedal desert rodent.

Chasing rodents across the desert had always been fun during my undergraduate days, but I was older now and there was nothing fun about chasing jerboas across the Iranian Desert. They are much faster than kangaroo rats. They live in areas that are almost free of vegetation and where a crust of pebbles overlays the sand below. When I began to chase the first animal I thought it would be like the old days, a bit of sport and nothing more. It was nothing of the sort. The jerboas had blazing speed and ran for long distances before diving into a burrow. As soon as I chased the first one I realized that things were very different in this desert. These rodents were highly specialized for escaping predators by exercising both great speed and superb agility. The kangaroo rat's escape strategy paled by comparison. I chased many animals, but was able to catch only one.

While I was chasing the jerboas, I saw a Ruppell's fox *(Vulpes ruppelli)*, a species common across Eurasia's arid belt. I also caught a long-eared hedgehog *(Hemiechinus auritus)*, an animal that epitomizes the great differences between the Iranian Desert and the deserts of the New World. This odd, fairly robust animal, weighing a half to three-quarters of a pound, is covered with stiff spines, shorter than a porcupine's but more abundant; it has long ears, a long snout, and is an insectivore distantly related to shrews and moles. It waddles around the desert in search of invertebrates and small vertebrates.

When I prepared the one jerboa I had caught, I found that its stomach was full of green plant matter, not seeds, a hint of a major difference between jerboas and the seed-eating kangaroo rats. Another difference is that jerboas may come out to feed during daylight, something kangaroo rats never do. Given the ability of the jerboas to avoid predation through their blinding speed, broken-field running, and sudden disappearance into almost invisible burrows (they lack mounds or other surface signs of their existence), I would guess that even in daylight it would be difficult for a predator to capture one of them.

During my time in Iran, I developed a good feel for the habitats and fauna of the desert. I found that small mammals were abundant, even though not many species lived in any one habitat. From the point of view of abundance, Iran and Arizona were quite similar, and both were markedly different from the Argentine Monte, where the scarcity of rodents had frustrated me time and again. The number of species found together is also an important attribute of a habitat, however. Across the sweep of arid North America, the number of species varies with rainfall—drier habitats support fewer species than wetter desert habitats. The number of species of small mammals in the desert habitats of Iran was comparable to the number found in North American desert habitats. The Monte Desert of Argentina was different, however. Its habitats supported fewer species of small mammals than either of the northern deserts. Again, Iran and Arizona were more alike than were Arizona and Argentina.

When my research ended, I decided I would return to Iran for extended fieldwork, but that was not to be. It was evident that there was great fear among the people of Iran. Some estimates said that one of every seven Iranians was a spy for the Shah, whose power was threatened by religious fundamentalists. The Shah, allying himself with the United States, had tried to modernize the country. But his popularity was slipping and change was in the air. Some people would talk to me only after I convinced them that I was not a CIA agent, which is not an easy thing to do. I am often suspected of working for the CIA, since my "cover story"—collecting rats—seems too far-fetched to be true. From conversations with Iranians who were willing to discuss the situation, I sensed that the rampant fear and mistrust of the Shah would lead to the overthrow of his government. In 1979, four years after I left Iran, the U.S. embassy was invaded and the Americans inside were taken hostage. The Shah fell, and the Ayatollah Ruhollah Khomeini assumed power. Desert research in Iran was no longer possible.

I was able to use the numerical data that I had collected in the field and from specimens in museums in Tehran and London to analyze the fauna of the Iranian desert and compare my findings with similar data from the Sonoran and Monte deserts. I found that when a large number of characteristics were considered together in a single mathematical analysis, the results supported my feelings in the field that the two northern deserts were more similar to each other than either was to the Monte. This is what I had expected to find if time was a major player in determining whether or not organisms evolved in similar ways toward desert specialization. Moreover, the kangaroo rats and the jerboas had each been associated with a desert for a long period of time, and both had become bipedal desert specialists.

Further analysis showed that there were also major differences between the two northern deserts. It had for years been assumed that jerboas and kangaroo rats were exact ecological analogues because they looked so much alike. Both were thought to be bipedal seed specialists. Some studies had even suggested that these animals were bipedal so that they could become more efficient seed eaters. My analyses of dental traits suggested that jerboas and kangaroo rats fill very different niches. Both are bipedal, but jerboas feed mainly on green plants, roots, and tubers, not on seeds, whereas kangaroo rats primarily eat seeds.

I also showed that the jirds are burrowing seed eaters. Their bodies are similar in shape to those of gophers or tucu-tucos, but their food habits are like those of kangaroo rats. Indeed, described in dietary terms, jerboas, appear to be bipedal gophers, whereas jirds seem to be burrowing kangaroo rats. I termed this phenomenon "niche-switch," because major portions of the adaptive traits of the animals in question seemed to have developed independently of other major groups of characteristics. Externally, an animal might appear to be similar ecologically to another animal in another desert (the kangaroo rat and the jerboa, or the jird and the gopher), but dietarily the animals that looked alike were not similar at all.

There were no gophers in Iran—no burrowing plant eaters like those found in both the Sonoran and the Monte deserts. But the jerboas that hopped around like little kangaroos ate the same food as gophers. There were no bipedal, kangaroo-rat–like seed specialists in Iran, but the burrowing jirds ate seeds. Convergent evolution was obviously a more complex phenomenon than had been assumed, but it had nevertheless occurred. To put it in different terms, I could predict many adaptations that would appear among unrelated desert rodents—thus faunal convergence was evident—but I could not predict whether or not exact equivalents would evolve in any particular desert.

This point is made clear by a consideration of the gopher niche. The two most functionally similar animals in any of the deserts I had examined, and two of the most distantly related, are the gophers of North America and the tucu-tucos of Argentina. They look alike, being hard to distinguish even if you hold them in your hands. They construct similar burrows and mounds, and both feed on green vegetation, roots, and tubers. A better example of a one-to-one pairing is hard to find. You might expect that this "gopher" niche would be similar in all deserts. But any thoughts I had along these lines were put to rest when I examined the Iranian Desert, which clearly supported a root- and tuber-eating rodent, but one that looks and moves like a kangaroo rat. The more I learned, the more complicated the story became of how life develops in a desert.

Impenetrable Land of Thorns

We began traveling at dawn. We crossed nine miles of prairie and entered an ancient track made by Indians that was widened by our laborers . . . We crossed a sparse woodland of thorns about six miles long, then a small clearing, then again into the interminable forest for another fifteen miles. The animals can no longer move due to the heat and we are exhausted by thirst . . . We arrived at six. The animals, almost dead from thirst, stumbled to the water, the packs still on their backs . . . The night begins and millions of mosquitoes assault us. We ate a horrible soup of boiled jerky with a few grains of corn.

Emilio Budin on collecting mammals in the Chaco, 1906, in Rubén Barquez, "Viajes de Emilio Budin: la expedición al Chaco, 1906–1907"

I received a Fulbright grant in 1976 to return to Argentina for six months to study the mammals of Salta, a large province bordered by five other provinces as well as by Paraguay, Bolivia, and Chile. Fulbright grants were primarily designed for people doing library research, so there were no funds for vehicles, equipment, and other expenses associated with fieldwork. Nevertheless, I thought that if I could get to Salta my students and I could live frugally and use our limited funds for fieldwork.

The Fulbright office presented me with a valuable gift, one that may have helped save our lives. It was a small plastic identification card that said that I was an esteemed guest of the president and government of Argentina. Police officers, the military, customs officers, and all government employees were asked to assist me with my work, which was of great importance to the nation. The card smoothed the way through bureaucratic tangles, and instilled confidence in nervous policemen and soldiers that I was not a threat to the peace and stability of Argentina. It would prove to be particularly important during this visit, for we had arrived at the height of what came to be called the "Dirty War." Salta province was a center of government extinction programs, especially in the outlying areas, where guerrillas were active and where we would be trapping.

The first problem I faced was the usual one of a field biologist: finding a vehicle. Buying one was out of the question, because the average vehicle cost more than my yearly salary in the United States. Renting one was also out of the question, because the weekly rental exceeded my entire Fulbright stipend. One of my students had a relative who owned an old truck that we could rent. We knew of no other vehicles available that could be rented for six months, so we closed the deal.

This truck could have been a national symbol of a country where how things look is more important than how well they function. It appeared to be equipped to explore the most remote corners of the province, but was beset by problems that required constant repairs. The truck showed its personality on the first trip. As we arrived in the city of Salta, the rear axle divided in two and the rear wheel separated from the truck. Only minutes earlier, we had been traveling at more than 70 miles per hour. Had this mishap not occurred in town we would probably have been killed. Characteristically, the truck that frequently seemed intent on killing us shied away from finishing the task. With this vehicle we intended to study one of the most challenging habitats in Argentina, the great thorn forest.

In Spanish, the Chaco is called El Impenetrable, the impenetrable forest. We found that the thorn forest daily taught us lessons in humility. Over several decades in the field I have worked in many habitats throughout the world, conducting research in blazing heat and frigid cold. Rain has pounded my field sites, violent thunderstorms have destroyed my camps, and howling winds have ripped the tents to shreds—with me in them. But of all of the places I have worked, none compares to the Chaco. It has brought me and my crews to our knees and made us aware of our limitations. It is a harsh land where most of the plants bear thorns. There are cacti, including some that become large trees that look like huge blue-green candelabras with enormous numbers of spiny branches. The shrubs also have thorns, and so do the herbs. Some of the thorns are of heroic proportions. One mesquite tree, the vinal (Prosopis ruscifolia), has thorns that are 10 inches long—remarkably tough, stiff spines that can slice all the way through a tire.

People graze cattle in the Chaco, and the gauchos who tend the livestock wear heavy leather gear for protection. Their horses also wear a wide apron of hardened leather, called a guardamonte, that extends a couple of feet outward on either side of the horse. The gauchos become leather-clad knights dressed to do battle with the thorn forest. The most deadly thorn is that of the vinal. There are one-eyed gauchos and one-eyed horses in the Chaco that became that way after encounters with the stiff spines, which are almost impossible to see if viewed head on. If you were racing through the Chaco on horseback and caught one of the spines full-force, it could penetrate the skull, leaving you impaled on the tree.

I visited parts of the Chaco during my first two years in Argentina (1971–1972) and again in 1974, but I did not work there extensively until 1976. My students and I wanted to study the mammals of the area because little research had been done there. We were poorly prepared for the difficult work, although in 1974 we had touched the edges of the vast region. Our vehicle then was a tiny Citroen truck into which were jammed traps, tents, rifles, stove, dishes, supplies, water, and the other accouterments of a field trip. Three of us traveled together up front, and one of my students had to sit in back with the equipment packed in around him. After the truck was crammed full we carefully stuck in two guitars and a bombo, a huge drum that made most of our trips with us and that permitted us to pretend that we were capable of singing folk music. Our trips would last for weeks, until we ran out money, gas, or water.

Gauchos' horses equipped with hardened leather *guardamontes* in preparation for entering the great thorn forest of the Chaco in eastern Salta Province, Argentina. (Photo by M. A. Mares.)

Our only major trip to explore the western Chaco in 1974 involved seven people traveling in two trucks, neither with four-wheel drive. We attempted to enter the thorn forest along the Tarija River, which forms the border between Argentina and Bolivia. The Chaco is a flat plain with only a slight change in elevation over a distance of six hundred miles. The soils are hardpan and when it rains the ground becomes impassable. Our trip coincided with the earliest hints of the rainy season. As we wound our way into the thorn forest the roads became increasingly treacherous, and within an hour first one then both vehicles got stuck in the muddy road, bogged down in a slick sludge that coated the tires and filled the treads.

We labored to free the Citroen, which we managed to do after working all day. We made camp where the other truck remained stuck. We would wait until the road dried before digging it out. Two days later we heard the distant sound of an approaching vehicle. There was no way around our truck, so we knew that whatever kind of vehicle it was—surely a four-wheel-drive truck at the least—the driver would pull us out.

Minutes later, a huge tractor appeared, its rear wheels more than 6 feet in diameter. We explained our situation to the driver. He used a chain to pull our truck from the mire. Asked about the condition of the road ahead, he replied that he had been stuck for five days. He did not recommend that we tackle the Chaco during the rainy season with our pathetic little trucks. The Chaco had effortlessly chased us away after we had penetrated its perimeter by only a few miles. In 1976 the memory of this experience led my crew and me to attempt our first field trips into the heart of the thorn forest in eastern Salta. We were slow learners.

We knew from the work of mammalogists over the previous sixty years that the region was rich in mammals. It abuts several major habitats, including the vast grasslands of central Brazil—grasslands that share many species with the Amazon rainforest to the north and the Atlantic rainforest to the east. The Chaco is the "mother habitat" of the Monte Desert, for during the wetter periods of the Pleistocene the Chaco extended into the area that today supports the Monte Desert. As the thorn forest retreated it left behind many plants that were adapted to aridity.

Earlier in 1976 the Chacoan peccary had been discovered in Paraguay, its description in *Science* magazine creating a stir in the zoological world. Later that year the species was reported from Argentina by Claes Olrog, and by my students Dickie Ojeda and Rubén Barquez. The discovery of a 100-pound animal thought to have been extinct for 10,000 years was surprising to many, but not to Dr. Olrog. He had heard of this unusual peccary for years, but had never been able to collect one. At the time, the small mammals of the Chaco were much more poorly known than the larger species, and not a great deal was known about them despite the discovery of the peccary. Indeed, before our work no detailed systematic survey of the area had ever been done. We were determined to spend a significant portion of time in the field in the hopes of clarifying the patterns of distribution of the mammals of the Chaco.

Research in 1976 began inauspiciously as vehicle and money problems soon manifested themselves. On our first trip we learned that the truck consumed gas at an extraordinary rate. Carrying extra gas in a 5-gallon can was futile. Such a small amount would only let us drive another 20 or 25 miles, a pitiful distance in the vast Chaco. We purchased a 60-gallon barrel and two 5-gallon cans. Now we had 70 gallons of fuel in the bed of the truck, in addition to the 30 gallons in the main fuel tank—100 gallons of gasoline would permit us to reach the most remote corners of the province. We were a rolling bomb.

Our work began on September 2, 1976, and at first we caught few mammals. By September 6 we had collected only sixteen specimens and were having problems with the truck. We decided to spend a night in a hotel in Tartagal, a town just outside the Chaco, and reenter the thorn forest early the next morning. By staying close to the jumping-off point we would be able to make good time the next day.

As I wrote in my field notes: "We got to Tartagal at 4:00 P.M. and got a room at the Residencial Premier. I have bronchitis and I am chilled. It's frigid out. We rested, ate, and sacked out. Then Rubén discovered something. He was covered with ticks and fleas. We all began to search and sure enough, we all had loads of fleas hopping on us and buried in the skin. I had a few ticks too. We can't figure out where we got fleas, but we have a pile of them. We picked fleas off for about an hour, then got tired of it."

In the morning, we left Tartagal and headed into the Chaco on dirt tracks. Our maps were inadequate. Often the roads appeared as dotted lines, indicating trails. Worse, some roads appeared as fairly good roads on the map, but did not exist. Occasionally we would come upon apparently good roads that were not even on the map. Once we crashed through thick thorn scrub only to encounter a paved highway in pristine condition crossing our path. It was two miles long and began and ended in thorn forest, almost as if had dropped from the sky. We never discovered its purpose, but drove over its smooth surface a couple of times for sheer enjoyment before plunging back into the scrub on the rugged dirt track.

The day was dismally overcast, threatening rain, which we knew would stop us until the roads dried. Our compass stopped working. We were left to navigate by maps that bore only a weak correlation to reality in an area where flat tires and car trouble could be deadly. I announced that I had an unerring sense of direction. I do not know why I said this, since my sense of direction is not especially outstanding. I thought we could certainly navigate the 120 miles that we had to cross to reach eastern Salta Province. After all, we had maps.

We headed northeast into the thorn forest. During the long day we saw no vehicles and passed no dwellings. We saw no people at all and spotted only a few animals. On we drove, hour after hour. The roads were very rough and I feared the truck would develop problems, since it was prone to do so at the most inopportune times. A few hours into the trip a problem did manifest itself. Dickie and Rubén, like most Argentines, were heavy smokers. Traveling with them meant that the air was constantly filled with tobacco smoke. Like all Argentines they would put out cigarette butts on the floor of a home, or the floor of a truck, by stepping on the glowing butt. We were traveling over a particularly rough patch of road when I asked them if they smelled gas.

"No," they answered.

"Well I do," I said.

I looked down and saw that gas was pouring across the floor of the truck. Both of them were still smoking contentedly. I told them to throw their cigarettes out the window immediately. They saw the gas rippling across the floor and quickly got rid of the cigarettes. We stopped the truck to assess the situation. The rough ride had caused the gas tank behind the seat to split open, allowing gas to leak into the cab. We were losing gas rapidly and were a long way from civilization. What to do? I suggested that we all chew gum, get a bar of soap, dig up some tape, and try to repair the tank. We were able to jam gum, soap, and toothpicks into the crack and then seal it with a bandage. The leak stopped. Now what? Go back and get the tank fixed? Return over three hours of rutted tracks to Tartagal? Never.

We forged ahead under the lowering skies, hoping that we would have enough gas to return to the city of Salta. Unfortunately, after driving ten hours in a "northeasterly" direction, we thought we were hallucinating when we saw dim green shapes appear in ghostlike fashion through the low clouds. Emerald mountains lay dead ahead. We knew that there are no mountains east of Tartagal until one reaches easternmost Argentina, six hundred miles away. Nevertheless, we were heading toward mountains. How could this be?

We stared at the deep green peaks with a sense of wonder. These were not the dry hills one might find in parts of the Chaco to the south. These were lush mountains with a verdant mantle of forest draped on their slopes, the emerald color even more striking when compared with the muted brown tones that had surrounded us all day. Such was my surprise that for a moment I thought we might have discovered a new mountain range. It was a mad idea of course. Maybe such things still happen in trackless Brazil, but not in developed Argentina.

It gradually dawned on us to our chagrin that we had made an enormous loop through the trackless scrub, burning our precious fuel and risking our lives. We had, by dint of my unerring internal compass, managed to arrive at a spot that was only eighteen miles up the road from where we had started in the dim light of dawn ten hours earlier. Our average speed from where we left that morning to where we made camp that night was 1.8 miles per hour, about the speed of an oxcart.

The Chaco had chewed us up and spit us out. Now we were low on gas, low on money, exhausted from the long, nerve-wracking drive, and mortified at our stupidity. We had no one to blame but ourselves. Idiots! We decided to spend a cold, wet, miserable night where we were and wait for morning.

In the morning we used almost the last of our money to reload the 60 gallons of gas we had burned on our hopeless hegira. Once again we headed into the Chaco. This time it would be Santa María at the eastern limits of Salta Province or bust. We had only $20 between us and there would be no further purchases until we returned to Salta.

Since we had no more money for food, over the next week we caught most of what we ate—rodents, birds, armadillos. This time we entered the Chaco via the Aguaray road, which was more of a road than a track. Several people had given us approximate kilometer distances to bifurcations, trifurcations, and intersections. Unfortunately, the odometer of the truck did not work. We could theoretically have estimated our position using the time-and-distance technique—so many kilometers an hour for so many hours would provide an approximate distance traveled. But alas, the speedometer did not work either.

We used every clue we could muster to decide where to turn. Often we stopped and debated which road to take. The sun was out so we were able to hold to an easterly course. After a few hours we hit an especially rough patch of road and again smelled gas. This time it was not the tank in the truck that was leaking. It never leaked again thanks to the gum. Instead, a Coke bottle had broken in the bed of the truck and glass had punctured the 5-gallon plastic container. Gas was pouring over our equipment. Now fuel would be very tight.

Late in the afternoon, after more than eight hours of driving, we drove into a tiny, dust-caked town. Through some miracle of navigation we had actually reached Santa María. Hundreds of parrots were flying about, screeching wildly: turquoise-fronted amazons (*Amazona aestiva*) with bright yellow heads, burrowing parrots (*Cyanoliseus patagonus*) with fiery red bellies, blue wings, and olive backs, and tiny, colonial-nesting monk parakeets (*Myiopsitta monachus*). Only a few dozen people lived there, a few whites and several Mataco and Toba Indians. As we drove to the outskirts of town to make camp, we saw many Chacoan cavies (*Pediolagus salinicola*), large rodents, weighing 5 pounds each, that resemble a jackrabbit with short ears and are close relatives of the fleet-footed Patagonian hare.

I had been unable to get a permit to carry a gun, although I had been told that the police would eventually give me one. It was especially difficult to get permission to carry a gun in the midst of the internal war that was under way. Carrying a gun was risky. We had brought an air rifle that we used to collect food. I decided to use it to collect a couple of the cavies, since we could get close to them. I shot two and we not only had fine specimens, but dinner as well.

After making camp we realized there were yet further difficulties to be overcome in the Chaco. The soil in the great thorn forest is so hard packed

The trackless Chaco near Santa María, Salta Province, Argentina. (Photo by M. A. Mares.)

that no tracks show on it. The forest is also unnervingly similar in every direction. Since the land is absolutely flat there is no sense of one part of the forest differing from another, no hills or mountains in the distance to offer a fixed point of reference. All the trees look the same.

As I began setting my traps, I immediately recognized that I could easily get lost. Even within 50 feet or our tents, it was difficult to say where I was in relation to camp. The sun was low and clouds had reappeared, making it difficult to retain any sense of direction. I returned to camp and decided to set my trap markers (brightly colored plastic flags) so that I would be able to see them from either direction. I even set extra markers. I ran out a line of traps over a distance of three quarters of a mile and then began following the flags back to camp.

I found about twenty flags but then could not find the next one. I moved in a circle around the last marker I had found, but to no avail. Where was the next one? I moved in ever increasing circles, always keeping the last flag in sight, but one part of the Chaco looked exactly like every other. No tracks had been left on the hard ground, which was as impervious to footprints as ceramic tile. I did not know how to get back to camp. The afternoon light was starting to wane, but I was too embarrassed to yell for my students.

I sat down next to the flag that had become my home base and decided to ponder my situation. I was not really lost, but had merely misplaced the camp. I could be no more than half a mile or so from where I wanted to be. As I sat there, I heard a sound. There were so few people around that the sound could only have come from the camp. I waited, then heard it again. I moved toward it, and after walking no more than 50 feet I located my line of markers again. Gradually and carefully, like Hansel and Gretel wending their way through the Black Forest, I followed the markers back to camp. I entered camp as if nothing was amiss, but I could not fool myself. The Chaco had given me a warning.

My students and I set 130 traps the first night and 301 the second night. For our efforts we caught only 4 leaf-eared mice (*Graomys griseoflavus*), a common species that occurs throughout Argentina. Had we set the traps in the dry forests of the American Southwest, we could have caught as many as 200 animals. The Chaco was a tough place to collect mammals. We could have caught the same species we had worked so hard to catch on the hillside near my house in Salta. For all of the effort, money, and time put into our first Chaco expedition, we had caught only 41 animals out of 1,127 traps set, a trap success of less than 4 percent. We had worked very hard for very little, which I would find was typical of research in the Chaco. We were low on gas, money, and food. We decided to return to Salta.

The brakes, the front end, and the U-joint broke down on the return trip. After repairing them we headed back to Santa María. In the interim, the weather had changed from cold and cloudy to blazing hot and sunny. Now water would become as important as gas in limiting our movements. My notes record, "It is hot today [September 27, 1976] (103° in the shade at 2:00 P.M.). We're lying under a tarp. This is a bravo [fierce] forest and there's no doubt about it. Because of the green vegetation, humidity is high (and the temperature is now 104°), so it's tough sitting here. To top everything else off, there are a jillion little bees and ticks that are bothering the hell out of us. There's no shade around, so all we can do is sit here. Jeez."

On the twenty-eighth, I wrote: "We are 20/535 total for this trip [20 animals out of 535 traps set, a rough measure of animal abundance]. It was hot yesterday (110° maximum) and hot today (109° maximum). We moved our locale. The bugs were driving us crazy . . . The heat is oppressive. It's not just the heat—110° is high but not oppressive, but the humidity is high. We are beaten down to nothing, weak, headaches. We've been very uncomfortable this trip—ticks, bees, ants, mosquitoes, lots of thorns. And above all, there is the heat, a stifling, oppressive heat."

On October 2 we again entered Santa María. But Santa María was not our original destination. We had intended to go to a small "town" called Santa Victoria that appeared on our map, but the road was difficult to find, and when we did find it (if it was indeed the right road), it proved impossible to use. There was supposedly another approach to Santa Victoria through Santa María, however, so we returned to that miserable settlement. The heat was intolerable, so after arriving we decided to see if either of the small stores had cold drinks, or at least had drinks that were cooler than the 100°-plus liquids that were in the truck. The store we entered was dark and cool compared to the broiling hot truck in which we had been confined for six hours.

We asked about cool drinks, and the owner, a white man, said he had cool beer. We were astounded and quickly ordered three large bottles. We were exhausted, having been in the field for more than a week in terrific heat, and were filthy. We looked like desperadoes. Rubén is garrulous and struck up a conversation with the old man. Since he was clearly not an Indian, Rubén asked him where he was from.

"I am from very far away, from a place you have not heard of," said the old man.

"I have traveled a bit," said Rubén, "where are you from?"

"I am from beyond Mars," the old man responded.

I was listening to the conversation with great interest. What an odd character we had encountered in this godforsaken place. He seemed to be in his seventies at least, heavy set, with a mustache and shaggy hair. His clothes were well worn and could have been homemade. He looked as if he had spent his entire life in the Chaco. Perhaps the hostile thorn forest that surrounded the town defined his whole existence. It appeared that too many years in this place had affected his mind.

"No, really, where are you from?" asked Rubén.

"Detroit," he replied.

What a crazy character, I thought, but now we have caught him in his lie. Can anyone from the United States have ever been in this backwater? Rubén returned to the table where the rest of us were sitting and we sipped our beers while we discussed the old codger. Perhaps just to be difficult, or to pass the hot afternoon hours in the darkened room, I decided to put him to the test. I walked up to the bar and greeted him in Spanish. He answered cordially. Then I said, in rapid English, "So, you're from Detroit, eh?"

He looked at me as if I had suddenly dropped from the sky in a beam of light. There was a hesitation, but I saw a dawning awareness in his eyes. He knew what I had said.

Then he spoke—in perfect, if hesitant, English.

"Yes, I was born there, but I haven't been there for many years." Like a Dr. Seuss character he said, "I am Sam."

Now it was my turn to be surprised. I asked him if he was a U.S. citizen.

"Yes," he replied, and after a long pause added, "but I lost my passport many years ago," again a long pause, "perhaps thirty or forty years ago."

"How did you end up in Santa María?" I asked.

He smiled ruefully.

"Yes, it is the end of the earth, isn't it?" He looked at me for a while, and then added, "You cannot go any lower than this."

He then related his story.

"My brother was with the Al Capone gang. In the 1920s he and Al Capone fought and formed rival gangs. Capone was going to kill my brother and the entire family. My father decided to take us as far away from the gangs as possible, so he brought us to Buenos Aires. That was more than fifty years ago. I was twelve at the time."

I thought about the family fleeing so long ago, traveling so far from the cold northern city to Buenos Aires. Buenos Aires, however, was as far removed from Santa María as Detroit was from the moon. I would guess that of the tens of millions of people who had ever arrived at or lived in Buenos Aires, few had traveled to Santa María—and only one had stayed there to live.

"It's still a long way from Buenos Aires to this place," I said. "How did you finally end up in Santa María?"

"Yes, it is a long way, isn't it?" Again the long pause, as if he was ruminating on his tangled past. "We moved. First to Córdoba, then to Tucumán, then to Salta."

I could see the paths of his life unwinding and leading him inexorably to this most dismal place buried in the great thorn forest.

"From Salta I moved to Tartagal. Then I came here." He stopped speaking for some time. I said nothing, still thinking about the great distance from cosmopolitan Buenos Aires to this place. It was a distance that could not be measured in miles.

He continued, "It was many years ago. I married an Indian and had a family. I will never leave this place."

"The gang wars are over in Chicago and Detroit," I said. "They have been over for more than forty years. Capone is dead. You need not have run for so long, or have run so far."

"Yes, this is the end of the world," he said as he gazed through the filthy panes of the windows toward the blazing thorn scrub outside. "I don't speak English here. I have not spoken it in some time. A North American came through here in 1950, I think. He left this."

He reached under the bar and pulled out a yellowed *Time* magazine from 1950, its thin pages crackling like parchment after more than a quarter century in the arid climate.

Again, he rooted around in the dusty darkness under the counter.

"Another American came here in the late nineteen fifties," he said, and out came another ancient edition of *Time*. "About twenty years ago a big Swede road into town on a horse. He was collecting birds. I do not remember his name, but he spoke English."

He was describing Dr. Claes Olrog, an ornithologist and mammalogist at the University of Tucumán.

"His name was Dr. Olrog," I said. "He is a friend of mine."

"Yes," said Sam, "that was his name. Is he still alive?"

"Yes, we saw him only a month ago. He is well."

Rustling around like a coatimundi in the thorn forest, Sam continued to find various other *Time* magazines, noting that every half decade or so a foreigner wandered through the area. Some were Anglican missionaries and brought books in English, as well as more recent magazines. He had kept in sporadic touch with the world and his native country through these rare visits, receiving a stroboscopic view of world events with bits of news filtering in once or twice a decade.

I discussed many things with him that day about life in the United States, but I could tell that Sam, in many ways as American as apple pie, had moved far beyond needing information about a country he had not seen in over five decades. Nevertheless, he was so pleased to see us that he pulled down from the wall behind the bar and gave to us the skin of a tamandua, the collared anteater *(Tamandua tetradactyla)*, a beautiful black and gold anteater that lives in the thorn scrub. He also offered us some bread that his wife had baked. We ate some bread, then bought the few loaves that remained, and took our leave.

Decades later I heard that Sam had died in the Chaco, no doubt buried in the hard-packed and trackless soil—safe from Al Capone at last.

After leaving Sam we decided to try again to get to Santa Victoria. This time we managed to reach the town, then drove to the nearby Indian village of La Merced before backtracking toward Santa Victoria to make camp several miles outside of town. We had seen many animals, including Chacoan cavies, brown brocket deer *(Mazama gouazoupira)*, and seriamas—tall, long-legged ground birds that run rapidly through the scrubland and eat a variety of animal and plant matter. They are in a family that is limited to the region and contains only two species, both of which occur in the Chaco. Since our previous visit, I had obtained a permit to carry a rifle in the car, so we were able to collect several rodents for dinner.

While Dickie remained in camp to cook the rodents, Rubén and I went in to Santa Victoria to inquire about possible collecting sites. We met an Indian who spoke some Spanish and lived in a Mataco village called San Luis, which was "near" Santa Victoria. He told us he was an armadillo hunter. He was going back to his village and gave us directions to his house so that we could visit him. His directions consisted of pointing into the Chaco and saying, "Go that way." We said we would travel to his village later in the day and would meet him there. He again pointed at a dusty track that disappeared into the thorn forest, indicating we should follow it.

At 4:30 P.M. we left camp in search of San Luis. We stopped in Santa Victoria, where the grocer in the tiny store told us that we would "never" find San Luis. But he was going there and, he said, we could follow him. It took him more than an hour to get ready to leave. At 6:00 P.M. we left for San Luis. He had a pickup with dual tires in back and blasted off through the thorn forest like a maniac determined to give us the slip.

As he sped away, enormous quantities of powdery dust blew up behind him, making it impossible to see where he had gone, or which of hundreds of intersecting animal and vehicle tracks he had chosen. Within three minutes we were lost, racing through the thorn scrub at 50 miles per hour, heedless of the millions of thorns surrounding us. My recent disorientation when trying to follow my trap flags was still fresh in my mind. Above the din of the bouncing truck, I shouted to Rubén that even if we found San Luis we would never find our way back to town.

It was stiflingly hot. Powdery dust billowed in through the closed windows, doors, and floor, coating us all over. The windshield was covered in dust and the air was opaque. We raced blindly toward the densest dust clouds, figuring the air would be most dust blown where the truck had passed. This blind pursuit continued for twenty minutes, during which time we never saw the truck, only different densities of dust.

If we stopped we knew that we would never find San Luis, but we also knew that if we continued trying to follow the elusive truck we would not be able to return to Santa Victoria. Suddenly, when we had just about given up hope, the Indian village loomed up through the dust. The mad grocer blasted on through the village, never stopping, and vanished. We were on our own.

The natives were incredibly poor and were clad only in rags. We could find only a few who knew more than a word or two of Spanish. In fact, they knew so little Spanish that we could hardly make the simplest concepts understood. We purchased a few armadillo shells (*Chaetophractus* and *Euphractus*) from some Indians who had piled them on the roof of their dwelling after eating the animals. We eventually found Señor Amaya, who was the only one in the village, apparently, who could speak even a few sentences in Spanish. He was the fellow we were looking for in the first place. We asked him about the rare Chacoan fairy armadillo (*Chlamyphorus retusus*), the beautiful chipmunk-sized armadillo that swims through sand. He said that this endangered species occurred in the area, but was so rare that he saw one only about every twenty years. We chatted for a while and then since it was getting dark I told Rubén that we should go back to Santa Victoria.

"Back?" asked Rubén, "Which way is back?"

Which way, indeed. We had driven a hundred yards into the village and neither of us could even find the point where we had entered it. Dozens of tracks converged on the village, each looking like the others. We made a complete circuit of the dusty village and realized that we had no idea how to return. We asked Señor Amaya, a man of few words, if he could point us in the right direction. He waved vaguely toward a distant corner of the village. We headed back into the scrub on one of the nameless ruts.

Soon more than a dozen tracks converged on ours and we were quickly disoriented. We traveled half a mile before we realized we were lost. We tried to return to the village, but could not find it, our futile attempt to backtrack only pushing us ever deeper into the hot thorn forest.

An uneasy feeling seeps into your entrails when you are lost in hostile territory. You realize that a whole series of small but injudicious decisions has gotten you into a life-threatening situation. First, we had violated my prime rule, which is not to go into the desert without lots of water. We had only one canteen between us. Without water, we would be in serious trouble when the daytime temperature exceeded 110°. We had also left without a compass, thinking that finding the village would be a simple matter. What to do? The various tracks we took got narrower, then the ground became muddy. We were forced to crash through mud puddles at high speed in order not to get stuck in the muck. Thorn trees scraped the truck mercilessly and the din in the hot cab was deafening.

Suddenly a Jeep appeared, moving fast along the narrow rut we were negotiating, but in the opposite direction. The Indian driving was clearly drunk, leaning precariously out of the topless Jeep. He had to stop as he approached us, though, for the track was not wide enough for two vehicles to pass. His eyes were bloodshot and unfocused. I asked him how to get to Santa Victoria. Thank God he spoke Spanish. He said he was heading there. Should we follow him? Yes. We *knew* we were lost but, drunk or not, he might not be. As we backed into the scrub to turn around, his Jeep zoomed forward and he was gone.

Incredibly, for the second time in the same day, we were again racing madly through the thorn forest, the speedometer moving above 50 miles per hour and the wayward Jeep nowhere in sight. Trees hurtled by and suddenly a trifurcation of the road appeared, each track mired in heavy dust. The dust seemed to hang more thickly down the left track, so off we charged, knowing that if we lost him we would have a very difficult night ahead. A mile or more down the track—another trifurcation of ruts. Screeching to a halt, we debated the choices.

"Which one?"

"The middle one."

Off we stormed into the forest. Then, suddenly, the dust cleared.

"Wrong rut!"

We backtracked and took a second rut. This time the dust was thicker. We were back on the "road," speeding after a drunk driver. He continued to widen the distance, since he knew the roads and did not have to drive through clouds of dust. I sideswiped a tree and hit a deep hole but the truck held together. Suddenly we blasted out of the dust cloud and were back in Santa Victoria, no longer adrift in a sea of thorns.

Now for the easy part—finding the camp. We made three futile attempts. Finally, we carefully measured the distance between each intersecting track (luckily I had written these distances down earlier, just in case we needed them), and at last found the one that led to our camp. Both of us arrived on the verge of nervous exhaustion. It took several hours and several glasses of wine for our adrenaline levels to return to normal. We slept little that night.

Dickie, Rubén, and I trapped in the area for a couple of days, then headed for a town called Balbuena, which we never found though we searched mightily. We came upon a place that on the map was called El Breal, but that consisted only of a single shack. Nearby was a marshy lake with many birds, including screamers, roseate spoonbills, herons, jaçanas, ducks, jabiru storks, and eagles. There is a scene in *True Grit* where a drunken Rooster Cogburn falls off his horse. He looks around helplessly at the place where he has fallen and says, "We'll make camp here." I looked at the lake, the waterbirds, and the surrounding vegetation, and thoughts of Balbuena, as elusive as El Dorado, vanished. "We'll make camp here," I said.

We spent the heat of the day splashing warm water covered with floating pig dung on ourselves and eating salt. It was the only way to keep cool. The wind howled all day, hot dust-laden blasts coating our wet bodies with a pale brown glaze. In the afternoon we trapped around the lake and in the surrounding thorn forest. By nightfall we were exhausted and tried to sleep with the hot wind roaring in our ears. The heat hardly diminished as the night wore on. Periodically, we splashed ourselves with water to cool down, but could get only brief relief from the heat. Late at night the winds became so strong that we were afraid that the giant mesquite tree under which we were camped—and which creaked ominously with each strong gust—would come crashing down on us, but it remained standing all night.

In the morning we explored the area. Plains vizcachas, large—10 to 15 pounds—rodents that live in colonies, were common and I collected one. We prepared the vizcacha as a study skin and a stew. Dickie and I spoke with the owner of the nearby shack, who offered to take us on an armadillo hunt

Dry thorn scrub near El Breal, Salta Province, Argentina. (Photo by M. A. Mares.)

that evening. Leaving Rubén to watch the camp, we picked up the man and his small dog shortly after dark and drove to a spot where armadillos were supposed to be common.

We had to drive very slowly because the hunter said he would get lost if we drove too fast. This concerned us, since we thought a local would be familiar with the area. But the man quickly explained: The endless scrub, which appeared to us to look the same wherever we looked, unscrolled before him in a recognizable pattern. But he had seldom been in a vehicle, and if the trees went by too fast he became disoriented. He could not get lost on foot, but could not find his way in a vehicle if it traveled faster than he could walk. We poked along to the hunting site.

When we reached it, a spot that appeared to us no different from thousands of square miles of surrounding scrubland, we immediately set off on foot. We walked swiftly and quietly for almost four hours, covering perhaps 15 miles. The moon was full and the hard-packed ground glistened white. The hunter and his dog were intent on finding armadillos. Dickie and I were in awe of being in the Chaco at night, following a man we did not know over terrain strange to us.

I told Dickie, "You know, if this guy has a heart attack, we're dead," since we had absolutely no idea where we were.

Around midnight we were getting tired and told our guide we were ready to head back.

"You are back," he said.

Sure enough, there in the moonlight about a hundred yards away was the truck. Dickie cynically said that we could have spent the entire night within a few hundred yards of the truck being led in circles just so the guy could impress us—a comment that said much more about us than it did about our guide. The hunter had impressed us greatly. But, unfortunately, we had found no armadillos.

A few nights later we were sitting around a large campfire after having trapped, prepared the specimens, and eaten some rodents. At about ten o'clock we heard an engine in the distance. It was odd to hear a vehicle coming from that direction, for it was not on the road to Santa Victoria, which was the only nearby town. In fact, it was not on the road to any town we knew about. Soon we saw a large truck heading slowly toward our camp. Eventually it stopped near us and the driver got out. Fear was written on his face. He had, he said, been lost for three days and thought he was going to die in the Chaco. He had been looking for Santa María, but had taken the wrong track. After wandering around a while looking for the right track (he did not have to explain to us, God knows), he realized he was lost. He had not seen a soul until he found us. He had had several flat tires, which he had fixed, but now he was low on fuel and food. We offered him rodent stew.

Remarkably, our little group, perhaps among the most lost people the Chaco had seen in some time, was to be his salvation. We even knew how to get to Santa Victoria, where he could get directions to Santa María. As he was leaving, he asked if we knew what he was delivering. We said no. He opened the back of the truck. Wine! In gratitude for our help and our food, he gave us a 5-liter demijohn of red wine before driving away. We quickly perked up and for a while as we drank the wine we laughed at the Chaco, briefly forgetting about the dust, the baking heat, and the headaches that awaited us tomorrow.

Over the years my students and I made a number of trips back to the Chaco. Each time we returned we found that the region was as challenging as ever. Once in southern Salta we made camp under a tree in a hot, dry area (of course). The night was warm, too warm for the mosquito netting, which kept out the slight breeze. After trapping, preparing specimens, and eating a dinner of sardines and crackers, we pulled out the guitars and the bombo and began singing the bittersweet songs of the Chalchaleros, Argentina's most famous and most traditional folk group. For hours we drank warm wine and sang, until drowsiness overtook us and we crawled into our sleeping bags to await the hot day that would soon be upon us.

Suddenly hundreds of kissing bugs (vinchucas) dropped on us from the tree, a veritable rain of biting bugs. Kissing bugs are reduviid bugs—also called assassin bugs—large, fat-bodied bugs with a long snout that they use to suck blood. Worse, they carry trypanosomiasis, or Chagas' disease, a parasitic infection caused by a protozoan that is related to African sleeping sickness. Chagas' disease causes heart failure and overall weakness, and can be fatal; there is no cure. It is very common in the Chaco—two thirds of the inhabitants are infected. It shortens your life and makes what life remains miserable.

These horrible bugs were now crawling all over us, invading our sleeping bags, landing in our hair, and furiously biting. We leaped off our cots and began swatting at the tough bugs, but their ferocious attack continued. I searched frantically for my mosquito netting so I could ward them off, while

Dickie and Rubén burrowed deeper into their stifling sleeping bags trying to escape the onslaught. After a frenzied fifteen minutes, the bugs left, disappearing into the thorn scrub as mysteriously and unexpectedly as they had appeared. We marveled at the assault for a while, then tried to sleep, but had trouble resting because we were worried that the bugs would return. But they did not and we never again experienced such an attack. Fortunately, none of us came down with Chagas' disease.

In 1990, during a survey of the mammals of Argentina, my students and I once again returned to the heart of the Chaco. Late one afternoon, in broiling heat, we came upon a group of ovens where charcoal was being made. Charcoal is one of the major wood products of Argentina, partly because many people heat their houses with it and partly because Argentines have an insatiable appetite for *asado*—grilled beef—which is always cooked over charcoal. The process of making charcoal is fascinating.

Charcoal ovens are domed semicircular structures made of bricks. Logs cut from the thorn trees are placed in the ovens and burned. Each oven has only one door, with a small vent at the back. As the wood burns it becomes carbonized—almost crystallized—the ideal wood for *asado*. The wood must not be allowed to burn completely, so the amount of air entering the oven is controlled. Incomplete combustion leads to interior levels of carbon monoxide that must be exceedingly high. The oven workers rush into the blazing hot structures and pull out the smoking charcoal so that it can cool.

It is an occupation that could have been devised by Beelzebub himself. The workers are black with charcoal dust, and their lungs must fill with charcoal and carbon monoxide as they drag the smoldering wood into the scorching sun. They wear leather gloves and a heavy apron, but have no other equipment. Oven temperatures must be in the hundreds of degrees. As the soot-covered workers race out of the glowing ovens, they surely find the intolerable heat of the Chaco refreshingly cool. I do not know how they survive and I do not know how long they last. If I had to vote for the worst job I

Charcoal ovens in the Chaco, Chaco Province, Argentina. (Photo by M. A. Mares)

ever saw, it would be making charcoal in the great thorn forest in the baking heat of summer.

The Chaco continued to dominate us on every visit. In my notes I wrote, "The oppressive heat, dust, wind, humidity, and bugs are beating us down. Mosquitoes and gnats at night; bees, wasps, biting flies, and flies during the day; spiders whenever. And the incessant heat—even at night . . . The ride was extremely hot, dry, and dusty. We were out of good drinking water. I realized I was in the early stages of heat stroke. As soon as we got to Pampa del Infierno (grassland of Hell!), we stopped at a gas station and guzzled cold Cokes, soda water, and mineral water. I found the salt shaker and ate a small handful of salt. The water and salt turned on my cooling mechanism and I began to sweat profusely. Within fifteen minutes I felt much better, so I told everyone to take salt and to drink as much liquid as possible. We had some difficulty getting water in town for the town itself is low on water."

On another day I wrote:

> I hate the Chaco. Always have. Nothing on this trip makes me change my mind. Today the heat is less oppressive, but the wind is

howling at 30–40 mph and the dust is swirling around us. I found an abandoned shed where I can get out of the wind while writing. We are going through water at a prodigious rate. The road was so bad yesterday that our built-in 15-gallon water tank split open, its precious cargo dribbling away as we drove. We salvaged about a half-gallon of it. However, it reduces our time here considerably. We almost certainly will have to go for more water within two days. Here I am in a hot, dusty, filthy shed with a mud-and-grass roof and holes in the wall that chickens walk through. Outside the dust blows and the wind gusts. The temperature moves inexorably upwards. THE CHACO SUCKS!

That day, when my field crew and I were camped at a ranch in western Chaco Province, was the hottest I have ever experienced. Our thermometer pegged at 120° in the shade, but the wind that blew was much hotter. I estimated that the shade temperature was over 130°. I deepened a hole that had been dug by a pig and lay down in it to try to escape the heat, burying myself as best I could in the hard soil. We all took turns trying to get out of the heat by using the wallow. We ate salt and drank hot water. Our skin was flushed, our heart rates were elevated. There was no escape. The wind seemed to make our ears crackle. Our lips split.

It was so hot that the birds began falling dead out of the trees. When the birds began to thud onto the hard-packed clay, we knew that this was no ordinary heat, even for the Chaco. These were desert birds, and their kind had persisted for thousands of centuries in the Chaco, despite the heat and aridity. I went up to a hummingbird that had fallen out of a tree. It was still alive, but panting profusely. I picked it up and gave it water. It recovered somewhat and I put it back on a branch that was in the shade of a tree. It looked as if it might pull through. All around us dead birds continued to fall onto the dry Chacoan floor. The heat was frightening.

In the late afternoon, when the temperature had dropped into the low triple digits, I told one of the students, "It's cooling off. I can feel it. The worst is over." I may even have believed it. Suddenly a blast of superheated air swept over us, almost making our skin sizzle. It burned our eyes and took our breath away. It continued to blow harder and harder, and hotter and hotter.

Neither of us said anything for a moment and then both of us started laughing. Our only defense was to laugh. The Chaco had thrown its best at us, but we were still alive and we could still laugh. That evening, a hunter who had been born in the Chaco told us that he had never experienced heat like that before and hoped never to again.

The next morning the heat broke and the temperature did not exceed 70°. The weather was cool and cloudy, but there was no rain. We used the day to regroup emotionally after the heat, to prepare specimens, and to write notes. One of the people living on the ranch shot a brown brocket deer and gave us half of it. We cooked deer empanadas that night.

One day the ranch caretaker asked if we could give him a ride to the house of a friend who lived "nearby" and whom he had not seen in two years (which made me wonder how "nearby" it actually was). He said there were many vizcachas at his friend's place and that we could collect some. We left in the late afternoon, following extremely narrow tracks through the scrub. We had to force our way through the thorn trees because the "road" was narrower than the truck. The muddy track required four-wheel drive. As we moved along the track, which was hardly wide enough for a person to walk along without getting scraped by the thorns, we crossed several other tracks.

"That one goes to Asuncion, Paraguay," the man would say, or "That one goes all the way to Brazil," referring to an extremely straight cut through the forest that went due north. Hardly anyone other than perhaps a person on horseback had gone over the trails in years, yet he knew where each one went. We crossed a dozen of these. Some went to nameless towns in the Chaco, others to an isolated ranch.

I thought about people entering one of these tunnel-like trails and knowing, somehow, when to take another track, or when to keep going on the original one. I remembered our armadillo-hunting friend from two decades before who crossed the Chaco at night and brought us back to a spot just a few feet from where we had begun. Such feats of navigation were astounding to me, but clearly were routine for the inhabitants of the Chaco.

After more than an hour spent traveling about 20 miles through the thorn forest, we reached the shack where the friend lived. The vizcachas refused to leave their burrows, but we did collect a skunk. Our guide decided to spend the night there and said he would return on horseback the next day. We headed back to camp on our own. We were alone in the Chaco at night.

The forest at midnight was even more confusing than when we had begun our journey. The narrow beams of our headlights could not penetrate the green wall of thorn trees on either side of us. We moved slowly through the mud and thorns, trying to decide which of two tracks to take. Then the engine died and the lights went out: total blackness and absolute silence.

We fiddled with the battery terminal and other wires whose function we did not know for about 15 minutes before realizing that the battery was dead. My great caution when dealing with the Chaco paid off, however, because I had brought along an extra battery. We connected it and the engine fired up. On the way back, we took two wrong turns—we were on our way to Brazil once and to Taco Pozo the second time—but each time we quickly realized our mistake and returned to the original track. Periodically we stopped to see if a passing vehicle had recently broken a branch. Any broken stems had to have been broken by us. We were carefully backtracking to camp. When we stumbled upon the camp at two o'clock in the morning I thought we still had 9 miles to go.

My last trip to the Chaco was in 1993. My students and I had tried unsuccessfully to find the Pilcomayo River, the dividing line between Argentina and Paraguay. On the face of it, that would seem to be a simple task, but we were in the Chaco, where things are never simple. Finding the river necessitated going to Formosa Province in the far north of Argentina. Once again we were among the desperately poor Toba Indians and lost on nameless tracks. No one spoke Spanish. We were reduced to saying "agua"—water—and making flowing signs with our hands. We must have seemed demented to the Indians, who did not understand the Spanish word for water. They were also unable to interpret our pathetic hand symbols for a flowing river.

Eventually, we thought we were told that the Pilcomayo River had gone underground—only isolated swamplands indicated where the river had once been. The actual river now lay much further to the north and could not be reached by vehicle, but only on foot or horseback. The Chaco had swallowed the river itself.

I wrote in my notes on that final trip to the Chaco:

> We left the Chaco on July 24. At about 5:30 P.M. we reached the pavement 6 miles east of Dragones. As we cheered the pavement and the end of the choking dust the engine exploded. A large chunk of the block flew away and a rod was left hanging out. A piston had blown down the highway. Dark was coming. We coasted to a stop and inspected the damage. We were staring helplessly at the ruined motor as a small, ancient taxi packed with six people suddenly appeared. I told Rubén he'd best go to Embarcación [the ancient port of Salta Province] to seek assistance. I told him to return next day and stuffed some cash in his hand for the trip. Within two minutes of a breakdown in the middle of nowhere Rubén, dressed only in shorts and a filthy T-shirt, was on his way to civilization.
>
> We made camp alongside the road. Within two hours the weather began to deteriorate. A cold front with strong winds blew in. It felt like Patagonia. Our tent began to blow away. I decided to relocate it to a small arroyo that offered some protection. There ensued a remarkably inefficient move of tent and equipment—the tent blowing into thorn bushes, the people slipping down embankments—before we were finally settled. I stayed awake in the roaring wind until 11:00 P.M., when Rubén arrived *walking*. The taxi had taken him to Embarcación (there was no help there), then on to Tartagal [city of fleas!]. The Automobile Club said they would send help in the morning. The taxi brought him back to within 400 yards of the truck, when the cab's axle broke and the rear wheel fell off!

We slept fitfully—four in the tent—with the biting wind howling all night. The next morning the Automobile Club arrived. We were towed to Tartagal, where the truck could not be repaired. The next morning we were towed 400 miles to Tucumán, arriving 11 hours later. The repairs would take three months. We left Tucumán on the bus for our base in Mendoza, 15 hours away. It was a frigid trip because the bus's heater was broken. Women begged the driver to do something, but he could do nothing but drive on through the black, cold, endless desert night. We arrived shivering and exhausted, with backaches and headaches. Two days later we were in Buenos Aires and thirty-six hours after that we were home.

All of our work in the Chaco in 1976 yielded fewer than 100 animals. The two trips in the 1990s brought in another 300 museum specimens. It was a small collection, but mammals are not abundant in the region. Of the specimens we collected, most were bats or rodents, and only a handful were marsupials, armadillos, or carnivores. Most were animals that were "common" in other parts of their range. But at least two, and perhaps four, were species that were either very rare or were previously unknown to science. We worked very hard to obtain each specimen and each animal has a story that goes with it. No doubt some of the specimens still have the dust of the Chaco on them if you look closely.

These specimens will some day be among the few specimens in collections that represent Chacoan species. At some time they will receive attention from taxonomists who will revise the taxonomy and systematics of the groups to which they belong, or examine the specimens for viruses or parasites, or utilize them in habitat preference studies, genetic studies, or biogeographic research. In the future some researcher sitting in a comfortable laboratory in an air-conditioned building will examine them, compare them to other specimens, and make scientific decisions based on the animals that we collected. Given current trends, that researcher may never have been in the field, his or her computer providing much of the information as to what is or is not a species.

When those researchers handle the animals, even if for just a moment, will they feel the stifling heat, the howling wind, the choking dust, and the vicious thorns? Will they feel the biting insects, see the desperately poor Indians, and taste the hot, salty drinking water? Can they ever know what it is like to see desert birds overcome by heat falling from the trees? I hope that somehow they will appreciate that these specimens were collected by field biologists willing to endure the harshest of conditions to seek out new life forms and learn new facts about animals in their native habitats. Although many of the specimens are of common species, I want those future researchers to know that *nothing* about the specimens themselves is common. They came from the Chaco.

The Devil's Town

You are always riding into the unexpected in these barren countries, stumbling upon strange phenomena, seeing strange sights.

John Van Dyke, *The Desert*, 1901

A paper I published in 1975 on convergent evolution came to the attention of Brazilian investigators who were interested in developing a research project on the mammals of the Caatinga (pronounced kah-CHING-ga), a vast scrubland in northeastern Brazil. The Caatinga is an odd habitat, one that at first glance appears to be similar to the Chaco, but that is much older and supports a different flora, with few species of plants being shared between the two regions. The Caatinga lies only a few degrees south of the equator but is semiarid for much of the year. The severe droughts that periodically afflict the region have led to nicknames such as the "Zone of Calamities" and the "Polygon of Drought." These refer not just to the climate, which is very unpredictable, but also to the horrendous effects of a variable climate on the people of the region.

The extreme droughts may last for a year or more and may occur only once every other decade. Their effects on people and livestock are immense. The Caatinga is one of the most impoverished areas in Brazil and also one of the most densely populated. For a decade or more the area may receive significant amounts of rain, becoming lush and green, an ideal land for crops. The climate during these periods is equitable, never too cold or too hot. Then a drought begins. Rain may not fall for months or even years, and the few showers that do occur do little to relieve the drought. Crops fail, ani-mals begin to die, and finally the people themselves face starvation. Mass mi-grations of people from the hinterlands to the few big cities that rim the Caatinga take place, with hundreds of thousands of people migrating into the cities in just a month or two. Chaos ensues.

When the Brazilian government developed the grandiose plan to build a major highway into the heart of the Amazon rainforest, one of its reasons was to lure poor people from the Caatinga to the Amazon with the promise of land. The idea was to curtail sudden population explosions in the major cit-ies of the Northeast when the Caatinga was hit by a long drought. The plan was not successful. The population density in the Caatinga hardly declined at all.

In my work in the Americas and Iran I had shown that desert mammals develop many predictable adaptations to life in arid lands. However, I had worked only in the temperate deserts that develop at about 30° north and south latitude owing to the global patterns of rainfall. In these areas deserts formed gradually because of slow increases in aridity that occurred over long periods of time. Habitat formation followed a repeatable pattern, with habi-tats changing over vast spans of time from tropical forest, through temperate forest, to grassland and scrubland, and finally to desert.

The mammals of the deserts also adapted gradually to aridity. Kangaroo rats and other heteromyid rodents, those paragons of desert adaptation, first developed in grasslands, as the earliest fossils show. Adaptations that equip animals to live in grasslands are similar to those required to live in a desert, although less pronounced. Animals sharpened these adaptations to new de-grees of precision as natural selection acted in the increasingly arid eco-system.

What would happen if a semiarid area was studied that was in the tropics and that had not undergone a gradual progression toward aridity? What would the mammals that lived there do if they could not adapt gradually to increasingly arid habitats? Would mammals even adapt to such an unusual area, and would they show adaptations to the semiarid climate? There were few such places on earth, but one was the Caatinga. As I thought about the advisability of conducting research in the region, it became clear that there were many unanswered questions about how the Caatinga affected small mammals. The area was old, so one might expect a highly specialized fauna, if age was an important factor in the development of a community of mammals. The region could harbor species that did not inhabit the wetter, tropical habitats that surrounded the region.

I went to Brazil in December 1975 to see whether a study of Caatinga mammals was feasible. From what I could determine from the literature, the mammals of the region had never been studied. I knew that if a long-term study were to be carried out there, it would have to be done by students. My teaching duties would no longer permit me to spend more than a month a year away from the university, and even that was granted grudgingly. For the first time in my life I would have to design and then entrust to graduate students a major investigation of mammal ecology and evolution in a region that had received almost no attention from mammalogists. This was not an easy step for me to take because I was accustomed to doing most of the field research myself.

After I arrived in Rio de Janeiro, I spent several days meeting with officials of the Brazilian National Academy of Sciences, the government research agency, and other organizations that were interested in a research project on the Caatinga. I was told that the government had allocated a significant amount of money to support diverse research projects in the area. Because of its great poverty, unpredictable droughts, and social problems, there was interest in studying the area's natural history. Diseases such as plague and leprosy were common there, and research on the biology of the mammals might prove helpful. Mammals were interesting because they had a direct impact on humans, either because people used them as sources of food or fur or because they carried plague and other diseases. Armadillos were known to suffer from leprosy.

I traveled to the Northeast with Dr. Aristides Leão, president of the Brazilian National Academy of Sciences and a wonderfully kind person. Dr. Leão had studied at Harvard, where he had befriended Joseph Kennedy, Jr. He had spent time visiting the Kennedys at Hyannisport and had even bounced the infant Ted Kennedy on his knee. After returning to Brazil, he obtained a position at the university in Rio. He was a laboratory scientist, although a naturalist at heart. He loved history and taught me much about Brazil, its people, and its habitats. We flew from Rio to Recife, a large coastal city in extreme northeastern Brazil, where we would procure a vehicle and head into the Caatinga.

I needed to determine whether my students and I could successfully carry out a major research project in the area, where it would be centered, and what the research design would be. I had to decide which species merited detailed study, where we would study them, and so on. Although this was a quick trip undertaken principally to choose the site of the study, I had brought along a few traps. I was confident that I would be able to decide if such a project was feasible. Whether or not students could do it was the question that concerned me. I had to decide if young graduate students could give up their lives in Pittsburgh (where I was a faculty member), move to a country whose language they did not speak, study mammals that were unfamiliar, and have the dedication to live and work there for two years.

We left Recife and I noted that the area had "many tropical elements, but also cacti . . . a weird combination." As we drove, stopping occasionally, I noted cavies (*Cavia aperea*), crab-eating foxes (*Cerdocyon thous*), armadillos, and a rock cavy (*Kerodon rupestris*) that occurred only in the isolated boulder piles and rocky hills that dotted the region. But "the most impressive thing is the number of towns, habitations, and people. Cacti are cut, hillsides burned."

We visited a private ranch near the town of Serra Talhada, a ranch where hunting was controlled. "The Caatinga here is thick and beautiful. It's disturbed, but not markedly so. I saw rock cavies living in boulder piles on boulders up to 15 feet high. Trapping is difficult. There are many thorn trees, stinging nettles, and dense scrub. One gets stung and scratched constantly. The rock cavies are probably primarily diurnal and apparently climb trees." I recorded that rock cavies were known to eat the bark of the *angico* tree, a local tree with a bark rich in tannins that was used to tan goat hides. I also noted that they ate the young leaves of *quixaba* trees, which they climbed. I found runways—well-worn paths—made by the rock cavies among the boulders.

Rock cavies were known almost exclusively from the Caatinga, so whatever we did in Brazil I knew we would have to study this species that seemed both threatened and specialized to life in the region. That same day I wrote:

> Right now I envision a study on rock cavy behavior and population ecology. It will be challenging, very difficult. I don't know how tough they'll be to trap. I don't know how easy they'll be to mark and spot. But it promises to be fascinating. The species is definitely in trouble with all of the hunting. Serra Talhada offers a good base for operations, a jumping-off point to the rest of the Caatinga. There are rock cavy areas nearby and two ice cream parlors in town. Today I am much more optimistic that a study could be done. I look across the thorny plains and see the distant bluffs, and it recalls many places I've been.
>
> This could be an exciting, though difficult, project. There are few mammals and lots of people. The specimens will be hard to get—it'll take a lot of driving and hacking with a machete. Is there a mammal fauna that has specialized for life in the Caatinga? I doubt it. Do the species here show special adaptations to aridity? Possibly. It's *very* dry. How about the rock cavy? The big enigma—apparently the only mammal largely limited to the Caatinga. How does it make it? Does it live with the ground cavy? [I had found a ground-dwelling cavy (*Cavia aperea*) that lived near, but not on, the same rocks as the rock cavy.] It's an intriguing problem—at least as difficult as my Monte study. But if we can pull it off it will clear up a whole major South American region—the Caatinga: enigmatic arid area of the tropics."

I was able to sample the fauna. I caught an unusual marsupial, the short-tailed opossum (*Monodelphis domestica*), a mouse opossum (*Marmosa karimii*), a forest mouse (*Bolomys lasiurus*)—a relative of grass mice (*Akodon*) that resembled them strongly—and a strange rodent that belonged to the spiny rat family (*Echimyidae*). This rodent is called a spiny rat because it has stiff hairs in the fur, though nowhere nearly as stiff as the spines of a porcupine. The animal was rare enough to be known only by its Brazilian name, *punaré*, and belonged to a genus in which it was the sole species *(Thrichomys apereoides)*, an indication of great age. "The *punaré* is a beautiful rodent— fine features and a long bushy tail. Really lovely. They live in boulder piles along with the short-tailed opossum." Later, I noted that "it looks so much like a packrat that it might as well eat cacti."

Also living in the Caatinga were animals not usually seen in a dry area, the small monkeys called white-tufted-ear marmosets *(Callithrix jacchus)*. Their primary food was sap and they obtained it by tapping the trees with their razor-sharp teeth to make the sap flow. Decidedly an odd place with an unusual fauna. Green habitats would persist for decades, then be succeeded by a year or more of drought, during which time the vegetation dried up and the people and animals died or migrated to wetter areas. The Caatinga had animals that were adapted to life in rock piles living alongside tiny monkeys that fed on tree sap. Marsupials, rodents, armadillos, and bats abounded, but there were almost no published works on the biology of any of the species. Indeed, there was not a single published paper on the mammals of the Caatinga, a region almost as large as New Mexico and Oklahoma combined.

We returned to Recife, where we would catch a plane to Rio. Christmas was only a few days away and Dr. Leão was eager to get home. I was just as eager to get back to Pittsburgh, where my three-year old and one-year-old sons were undoubtedly bursting with excitement over Daddy's return and the Christmas holiday. In Pittsburgh there would be twinkling lights and snow. In Recife there was blazing heat and humidity.

"I went for a walk in Recife. Most stores were closed. Christmas in Brazil is an amazing magnification of the commercialism of Christmas in the U.S. It's so different from Argentina. Garish 15-foot-high angels kneeling on plywood clouds and lit from within 'grace' each street corner. Strings of colored lights are everywhere. Stars, colored Christmas balls, artificial trees abound. It is mimicry carried to the extreme."

Any large city near the Caatinga, such as Recife with more than a million inhabitants, has enormous numbers of the extreme poor. These people are forced to beg to survive, for there are no social services available to them.

The contrast between the neon-lit downtown at Christmastime and the huddled poor along the bridge at the edge of the city was stark. After my walk in Recife, I noted:

> Beggars—Brazil has hordes of beggars, from children to ancient people. Each one is more pathetic than the preceding. One young girl was about 12. She had a pretty face and was in an advanced state of deforming leprosy. Her feet were being eroded away. One was only a tiny stump; one was halfway gone and swollen, pink, and horribly misshapen. The beggars line the bridges where people walk to take in the cool breezes or to see the Christmas lights of the city. One pair of beggars looked as if they had crawled out of a crypt or were about to crawl into one. An older woman, apparently the mother, had a gaunt stare and sunken eyes. She was reclining on the sidewalk cradling a young girl (6–8 years old) who had a lost and vacant look. Both were dressed in white shrouds and the girl had a wracking cough.
>
> Another old fellow was incredibly misshapen, skin very black, body rather bloated and twisted, legs merely small curled appendages, arms useless, bent protuberances. He was perhaps three-and-a-half feet tall, if that, and was wild-eyed and sitting in a beat-up wheel chair. Distress doesn't describe what one feels encountering these people.

I left Recife the next day and returned to the snows of Pennsylvania in time to celebrate the holidays with my family. The twisted beggars on the bridge in Recife were physically far away, but they persisted in my thoughts as a haunting reminder of a part of the world that most people seldom see.

After my return to the United States I wrote a proposal that was approved by the Brazilians; they promised two years of funding and three vehicles. An official at the Academy of Sciences called me and said that the Academy had found another town to use as home base. Instead of Serra Talhada, the nice

little town with the two ice cream parlors, the Academy had located an abandoned agricultural school in a rough-hewn town in an arid part of the
Caatinga. The town was named for the devil, Exu (pronounced eh-SHOO),
so the devil's town would be our new home. I could hardly send students to
live in a town named for the devil with habitats I had not seen. Thus I had to
return to Brazil in July of 1976, only a month before I was to begin my
Fulbright research project in Argentina on the mammals of the Chaco.

Caatinga scrub forest near Exu, Pernambuco, Brazil. The distant rocky uplands provide a moist microclimate that permits mammals to persist in the region during extended droughts. The granite on
the ground is called *lajeiro* and is common throughout the region, where it supports vegetation different from that in the rest of the Caatinga. (Photo by M. A. Mares.)

Once again I flew to Recife, then drove 14 hours to Exu, which is in the
state of Pernambuco, one of the poorest states in Brazil. I visited the abandoned agricultural school that would be home for the next two years. Parts of

it were falling down, but several gymnasiums were available for use in research, as well as several abandoned small houses. It was primitive, but it had electricity most of the time. I found it "absolutely fantastic," although I noted that the town itself was "rather primitive."

The town of Exu was famous for a feud between two families that had been going on for decades. Dozens of people had been killed, and every year one or two more were shot, stabbed, blown up by bombs, or otherwise eliminated in a seemingly endless exchange of hostilities. The two extended families—each with hundreds of distant relatives—were dead set on continuing to murder members of the opposing family. Sometimes the government declared martial law in an effort to control the killings and then the army would swoop down on the town. But these interventions were short-lived, and soon someone would be gunned down once again and the feud would continue.

There is always a risk in sending students into the field, where they must make innumerable adjustments to the daily challenges of field research. If they are living in a foreign country, they must deal with a new language, a new culture, and new laws. They are accustomed to electricity, clean water, houses with windows, vehicles that function, food that does not make you sick when you eat it, and so on. But in the field things are different. They may be far from medical care or the comforts of society that most of us take for granted. They may be treated wonderfully by the local people, but miserably by the police. They may be suspected of being CIA agents or they may be hated simply because they are Americans. I tell them that their human rights stop at the U.S. border. This is not exactly true, but it is good to remember how much we take for granted in America. Mostly, it *is* true.

Perhaps most important, students will have to learn to live together in the field. In a foreign country field researchers are thrown together in a situation that in many ways is as intimate as marriage. One eats, sleeps, visits, and works with the same people, day after day. Every quirk a person has (and we all have many) becomes irritating to the others. People may even come to hate each other. Being in a foreign country—being the only ones in a region who speak the same language and share a culture—draws the researchers together, whether they wish it or not, but such enforced togetherness can eventually lead to great animosity. Even married couples are often torn apart in such situations. Put three guys together, and things are even more difficult. To their credit, my students made it through a two-year field experience with only moderate neuroses and disagreements. But research in the Caatinga was not easy. It began rather shakily.

Shortly after their arrival in Exu one of the students walked into one of the many bathrooms at the agricultural station. There, much to his surprise, was a bat hanging on the ceiling in a corner of the small room. He carefully crept out of the bathroom so as not to scare the bat away. First question for an inexperienced field researcher: Should a solitary mammalogist weighing 175 pounds tangle with a bat that weighs half an ounce? The obvious answer was "No." It was time to call for help.

The three students huddled together to devise a strategy to capture their first Brazilian bat. Much planning was involved. Problems were discussed. Could the bat be rabid? Why else would it be sleeping in a bathroom in full daylight? Of course it was rabid! Special measures would be required. Each student rushed off to don protective clothing—heavy shirts, gloves, and other protective gear dug out of the boxes of field equipment they had brought from the United States. Long-handled nets were sought to keep the wily bat from fluttering away.

Fortunately during all of this hurried but fairly quiet activity the bat remained hanging in the corner of the bathroom, blissfully unaware of the well-protected threesome of field biologists about to descend upon it. Fully 500 pounds of mammalogists with more than a half century of formal education were ready to tackle the chiropteran featherweight that hung unperturbed in the little bathroom.

They stealthily crept back into the bathroom with gloves and nets at the ready. Hearts racing, they silently approached the bat, which still seemed unaware of their presence. As the three of them slunk toward it, the bat suddenly moved. One of the students shrieked, and all three ran pell-mell out of the bathroom and back into the laboratory, with much clattering of equipment. They closed the door behind them. It was time to review the plan again. This bat was not to be taken lightly. It could escape. It could even attack!

Once again they huddled. This time, they decided, only a full frontal assault would suffice. They would move swiftly and decisively before the creature with a brain the size of an insect could decide what to do. They charged the bat before it could attack them. Nets flew and the bat was theirs. It had not escaped. They had captured their first tropical bat!

In fact, the bat was dead. It was mummified and had not drawn a breath in a year or more. The years of fieldwork that lay ahead of them had begun, however inauspiciously.

Foreign research often entails unexpected problems. While giving a lecture at Harvard after I had returned from Argentina (about six months after the students had begun their work in Exu), I was told by a faculty member that he had heard that the students that I had sent to Brazil were primitive people. I said that all of them were big city boys from Pittsburgh, and while Pittsburgh is not New York or Paris, I would hardly call it primitive. Jonas Salk discovered the cure for polio there. The city has an excellent symphony, the Carnegie Museum, and other accouterments of sophisticated living, not to mention the Pirates and the Steelers. That my students would be described as primitive was especially surprising considering that they were living in Exu, which could rightly be described as the armpit of the Caatinga, itself one of the most primitive regions in South America. I asked for more information and he said that he had heard from a well-known Brazilian biologist that the students did not know how to use bathrooms and were doing their business on the porches of the buildings in which they were living.

How could such a rumor have gotten started? How could it be given any credence? I wrote a letter to the students telling them I had heard a rumor, at Harvard no less, that "you guys are shitting on the porches. What gives?"

They wrote back that since the agricultural school they were using for their research had been abandoned for many years before they arrived, the townspeople were in the habit of walking through the area and many of them relieved themselves on the porches. It was a problem that was diminishing with the presence of the students, who chased people away from the buildings because they did not want anything stolen or damaged. A Brazilian scientist had come by to visit them while they were out of town. Looking around, he noted the fecal matter on the porches of the dozen abandoned buildings and assumed it was those pesky North Americans who could just not adapt to modern toilets. His observations were dutifully reported to Harvard.

Exu was among the grimmest and most primitive towns in Brazil, a place where some of the inhabitants still believed the earth was flat. Since there is little formal education in poor towns like Exu, the people are subject to con men, hucksters, and others who make a living off mass ignorance and poverty. Like the snake oil salesmen of yore, they come to the small towns to prey on the ignorant and the superstitious.

One day Exu was visited by the "Profesor das Cobras"—the Professor of the Snakes—a con man who traveled with a box of nonpoisonous snakes and who sold a magic cure that would protect against the bite of poisonous snakes. The man would take out a boa constrictor, a nonpoisonous species, and allow it to bite him. He would then take a small candy tablet (which he had purchased in Recife and which was not sold in the interior) and eat it. Miraculously, his arm did not become red and swollen, and he did not die.

In Exu, there are many poisonous snakes, including several vipers, a species of large rattlesnake, and coral snakes. All are deadly, and it is not uncommon for people to die of snakebites or lose limbs when bitten. Everyone knew someone who had been bitten by a poisonous snake. Now the "Profesor" had arrived with a cure for the bite of any snake, including the deadly coral snake. Truly, people thought, this was a miracle, and they bought many candy pills.

It was the "Profesor's" bad luck that my three students were in Exu, and that a postdoctoral researcher, Dr. Laurie Vitt, had come to live with them for a year in the abandoned agricultural school. Vitt was a herpetologist studying the reptiles of the Caatinga and had been paying children and adults in Exu to bring him any lizards and snakes they found, while cautioning them to leave poisonous varieties alone. Now as he watched the "Profesor's" pitch he realized that children would think that the worthless sugar pills would protect them from poisonous snakebites.

Laurie and my students later found the con man having a drink in the local bar. Laurie sat down next to him and said he had heard about the pills that cured snakebite. The "Profesor" said, yes, the pills surely did the job. Laurie then asked if they would protect against the bite of any kind of snake. "Absolutely," replied the "Profesor."

"That's great," said Laurie, "I'll give you a thousand cruzeiros if you let this snake bite you." He then opened the bag he was carrying and showed the "Profesor das Cobras" a large and very deadly coral snake. The "Profesor" took one look at the snake and recognized it immediately.

"Ah, no," he said, "I cannot let the snake bite me for I just had an alcoholic drink and that causes the pills not to work for hours."

"We'll wait, and you can let it bite you later," said Laurie.

"Ah," said the "Profesor," "that will not be possible either, for I have had sex within the last month, and the pills do not work if you have had sex."

The excuses continued, so Laurie and the students trooped down to the police chief's office, where they explained that the "Profesor" was a con man who was selling worthless pills that could result in people dying from the bite of a poisonous snake. The chief and the four biologists went off to find the man. By the time they got to the bar, the "Profesor das Cobras" had packed his box of snakes and left Exu.

Some months later the "Profesor" appeared in Bodocó, a dismal town two hours south of Exu where one had to go to send or receive a telegram, and began—amidst much fanfare—to sell his miraculous pills. As he was beginning his pitch, someone told him that the North American biologists from Exu had just arrived. Without further ado the "Profesor" quickly packed his bags and left town, undoubtedly hoping to find a place that was more iso-

lated and, more important, one that lacked field biologists who could identify the snakes.

The work of my three students in the Caatinga provided a wealth of data on
the mammals of the area. In a series of six lengthy papers, Karl Streilein examined the population biology and behavior of most of the small mammals
of the region, showing how their populations responded to the unpredictable
droughts. Far from being specialized for aridity, most small mammals died
off during times of aridity. Only in small patches of habitat where moisture
was retained, such as the isolated rock piles and small rocky hills, did species
survive the droughts. My capture success rate with rodents in the Argentine
Monte Desert had been low, about one animal per 100 traps set, but capture
success in the Caatinga during a drought was pathetic, with only one animal
being captured for each 20,000 traps that were set. From these isolated frag-
ments of a population, the animals would increase rapidly during periods of
abundant rain. It was during such periods that plague and other diseases became common.

In the rock piles, immune in most ways to the prevailing droughts, the
rock cavies maintained stable populations throughout the year. Thomas
Lacher found the animals to be as difficult to capture as my initial foray had
predicted. In fact they were untrappable. He tried many types of traps (metal
live traps, mesh cage traps, wooden traps, snap traps) and a variety of baits,
but nothing would induce the animals to enter traps. If a student is dedicated
and tenacious, however, there is always a way. Tom noticed that rock cavies
raced deep into boulder piles or hollow trees at the first sign of danger. The
hollow trees provided the mechanism for their capture.

A rock cavy in a hollow tree retreats to the base of the tree, quite far from
the hole where it entered. Tom would reach into the tree, but usually the
hole was too small for his arm, or the animal was out of reach. He decided to
look for boys with thin, long arms and hired them to reach into the tree and
pull out the cavies. After a cavy ran into a tree, Tom drilled a hole in the base
of the tree just large enough for the boys to reach in and grab the rodent

A rock cavy *(Kerodon rupestris)*. (Photo by M. A. Mares.)

(which, fortunately, did not bite). Using this technique he was able to capture many animals and establish a colony in one of the abandoned buildings at the agricultural school. He brought in boulders and trees and constructed as natural a habitat as possible inside the gymnasium.

While this was going on in Exu, I attended a conference where a paper was presented on hyraxes, rodent-like animals related to elephants that live in boulder piles in dry thorn scrub habitat in central and southern Africa. The author showed a film about hyraxes and for a moment I could not believe my eyes. The film could have been shot in the Caatinga and the hyraxes could have been rock cavies.

The hyraxes are unusual in that they form harems, where a single male mates with a large number of females. The male is able to do this because the females need the rock piles in order to survive. Thus if a male can set up a territory that allows females to enter but excludes other males, he can develop a harem. It is a strategy similar to that used by the pashas and sheiks at the caravanserai in Iran, whose harems were lodged in the haremsarai. The sheik had the resources (wealth) to accrue the women, and the means (eunuchs) to keep other males away. So did hyraxes. I was certain that rock cavies did, too, and wrote immediately to Tom to alert him to this possibility.

His subsequent study showed that rock cavies defend rock piles and develop harems, exactly like rock hyraxes. In fact, cavies and hyraxes, which are about as distantly related as mammals can be, not only look alike, but are similar in almost all aspects of their reproduction, ecology, and behavior. It is one of the finest examples of convergent evolution discovered to date, involving as it does not only external morphology but fundamental functional biology and ecology as well. Later we extended this research to rock-dwelling mammals in general and found that there is a remarkable degree of convergent evolution that has developed among mammals that have specialized for life in isolated rock piles.

To live in rocks requires certain adaptations and these develop among all species. As a simple example, living on rocks necessitates feet that are specialized for gripping. There are only so many ways to do this, and several methods have appeared among the distantly related animals that have moved into rocks. In essence, their feet have converged on sneakers: various types of ridges, suction cups, and other structures have developed on the feet of all rock specialists. Most have harems.

The rocks also act as catchments for water, and green vegetation, including trees, is available in the rocky areas even during arid parts of the year. The animals develop the ability to climb into the trees, where they feed on green leaves. Animals in different orders or families that have specialized for life in rocks are more similar to each other in most characteristics than they are to species that are closely related to them. The force of natural selection in response to similar adaptive challenges has caused them to diverge from their relatives and converge toward each other.

My third student, Michael Willig, studied the bats of the Caatinga, carrying out one of the most extensive studies ever done on the morphology, ecology, and reproduction of tropical bats. It remains the only ecological study ever done on the bat fauna of the Caatinga. His extensive collection of bats will provide basic data for generations of mammalogists, for all are preserved in major museums and are accessible to scientists from throughout the world.

One species he studied was a most unusual little bat, the South American flat-headed bat *(Neoplatymops mattogrossensis)*, which dwells in rocks. This tiny bat lives between flakes of granite, has a flattened head and body, and wedges itself into the cracks in the rocks during the day. Its adaptations to life in the rocks even include the "tennis-shoe" pattern that characterizes so many rock-inhabiting mammals, but it is not the feet of the bat that are specialized for rocks. Instead, the tops of its forearms—the part of the body that comes in contact with the rocks as the bat moves between the sheets of granite—have leathery tubercles, little knobs much like those on the sole of a sneaker that it uses to provide traction on the slippery rock surface. The name *Neoplatymops* means the New World *Platymops*; in the same African rock piles where the hyraxes occur, the African flat-headed bat *(Platymops)* is also found.

Together my students and I learned that the Caatinga does not have a unique fauna. Like the Chaco to the south, the Caatinga supports a mix of species from surrounding habitats. Only one or two species are limited to the area. During periods of aridity most of the mammals either die out or retreat to moist habitats, such as the rocky areas with their green vegetation and water catchments. There are no arid-adapted mammals in the Caatinga, its droughts notwithstanding. The region, even though it is as old as any desert, is not a desert from the point of view of its mammals. Indeed, it appears to be even more challenging to mammals than a desert. Aridity is a difficult hurdle for a mammal to overcome, even if the evolutionary history of a species has taken place in habitats that became increasingly arid over long periods of time. But for tropical mammals adapted to warm, wet habitats throughout the year, the sporadic droughts that exceeded in length the life span of most small mammals were simply too much of a challenge for adaptations to aridity to develop.

Populations inhabiting the Caatinga would have only a year or so, perhaps only one or two generations, during which adaptations for an arid existence would be favored. Suddenly their offspring would face ten or twenty years of warm, wet weather, a period of time during which twenty to forty generations of small mammals would never experience arid conditions. The Caatinga's wild swings of climate made specialization by small mammals for the unlikely droughts all but impossible. We had documented convergent evolution, to be sure, but it was not to aridity. Instead it was to the unusual extruding piles of granite, a habitat that occurs sporadically around the world, yet one that demands adaptations of its mammals that are as stringent as those demanded by the driest desert.

When we left the Caatinga and published our research the region could no longer be considered poorly known, from the point of view of either its mammals or their ecology. We had taken a major biogeographic region and clarified its mammal fauna, ecologically, behaviorally, and biogeographically. Our research over the years has also left a wealth of scientific specimens of mammals in Brazil and Argentina that will be studied by future generations of mammalogists. Where before there was little to study in the way of extensive series of well-prepared specimens, our study specimens are now available to all students and scientists in those countries.

In the Shadow of the Pyramids

How many animals we have learned about for the first time in this age; how many are not known even now? Many things that are unknown to us the people of a coming age will know. Many discoveries are reserved for ages still to come, when memory of us will have been effaced. Our universe is a sorry little affair unless it has in it something for every age to investigate . . . This age will glimpse one of the secrets; the age which comes after us will glimpse another.

Seneca, *Naturales Quaestiones*, first century

The work in the Caatinga made it clear that the semiarid tropical woodland was not a suitable habitat in which to study mammals that were adapted to aridity. And more fieldwork in Iran was impossible because of the unstable political situation. But I was determined to find an Old World desert where long-term research could be done, and the greatest desert in the world is the Sahara. In August 1977, Duane Schlitter, curator at the Carnegie Museum in Pittsburgh and an expert on African mammals, and I made a reconnaissance trip to Egypt to determine whether or not funds could be obtained to study mammals of the Sahara Desert. This was my first trip to North Africa. My time in Egypt began inauspiciously since my bags did not reach Cairo until two weeks after I did. I spent those two weeks in summer in the Sahara Desert with only the clothes I wore on the trip to Cairo.

The day we arrived in Cairo, city of a thousand minarets, we were met by an Egyptian driver who had been hired by the U.S. Embassy. He whisked us through customs and then took us on a wild ride to the city. I quickly learned that in Cairo everyone drove while honking the horn; the din of Cairo's traffic was deafening. As we neared the vicinity of the embassy the driver informed us that the embassy was closed. He deposited us on a busy street and sped away, neglecting to tell us the location of the embassy or suggesting where we might go. Eventually we found the embassy and fortunately the science attaché was working late that weekend. He led us to a guesthouse.

On arrival, we received an invitation to attend a dinner at the home of Harry Hoogstraal, a famous parasitologist. Hoogstraal had been studying mammal parasites and their diseases in Africa for decades while working for the Naval American Medical Research Unit (NAMRU). We were exhausted after flying all night and most of the next day to reach Cairo and wanted to decline. But another man at the guesthouse had also been invited to the party and had hired a cab to take him to Hoogstraal's home on the outskirts of Cairo. We reluctantly agreed to accompany him.

The driver did not speak English. Worse, he had no understanding of Cairo. He battled the dense traffic for an hour before admitting he was lost. We found a business that was open and asked if anyone spoke English. A young man who spoke English knew the address we were seeking. He pointed out that we would never find it if left to our own devices or to those of our driver, so he got in his car and led us to Hoogstraal's home, a trip that took an hour. He would accept no money from us, saying that our gratitude was enough.

"Welcome to Cairo," he said, and drove away.

We spent a pleasant evening with Hoogstraal, whose home looked like a museum, being a spacious marble villa filled with works he had sculpted. We did not return to our rooms until after midnight.

In the morning we met with researchers who would work with us if a grant was forthcoming to study small mammals in the Sahara Desert near Cairo—Egypt's Western Desert. We met Dr. Kamal Wassif, the leading mammalogist in Egypt, who was interested in working with us. We also met Hoogstraal's field assistant, Ibrahim Helmy, an outstanding field mammal-

ogist who would sometimes conduct extended field trips into the Sahara using camels, rather than vehicles, and who, three years later, coauthored with
Dale Osborn the authoritative work *The Contemporary Land Mammals of
Egypt*. Helmy pointed out that the Sahara, long the site of major conflicts,
was a dangerous place to work. During our visit, Egypt and Libya were at
war. While examining habitats near El Alamein, the site of Field Marshal
Montgomery's victory over Rommel in World War II, I saw caravans of trucks
filled with wounded servicemen returning from the front.

Helmy showed us a photograph that he had taken. He had come upon
a shrub on a small sandy hillock. Skeletal legs with boots attached stuck
out from one side of the mound and a skull encased in a helmet protruded
from the other. The body had been covered by sand and a shrub had taken
root there. Dogtags were still around the neck and an overturned Jeep was
nearby. In the dry desert air, the Jeep was almost as fresh as the day it had
driven over the land mine that killed the driver, a British soldier who had
been missing since World War II. Helmy collected the dogtags and turned
them in at El Alamein, along with information about the location of the
skeleton.

He told us the story to impress upon us that the desert was still filled with
active land mines—they may explode a century or more after they were
laid—and that neither they nor the Sahara itself should be taken lightly.
Driving or walking where there are no paths or roads can be deadly if you
trip a mine. Helmy had worked in the Sahara for decades and knew what he
was talking about.

Some years later, Duane told me, Helmy was driving a group of biologists
through the southern Sinai. The track wound through a wadi, Arabic for
gully, that the team had traversed many times. But this time an unusually
heavy rain had fallen since their last trip, and the Jeep hit a land mine that
had been washed down from the slopes of the gully and then covered by the
wet silt and sand. Unfortunately, everyone in the Jeep except Helmy was
killed. He survived by some miracle, but lost three limbs in the explosion.
Although critically injured, he managed to crawl to a nearby main road,
where he had great difficulty getting any vehicle to stop to render assistance

because everyone is afraid to stop for any reason in such areas. Eventually, he was picked up, and recovered after a long convalescence.

Before we left for the field, Duane and I toured Cairo. I loved the filthy, bustling city, a vast sprawl of ancient dwellings along the Nile. We drove by the City of the Dead, an enormous cemetery with thousands of crypts that has been colonized by more than a hundred thousand of Cairo's destitute inhabitants. They live in and among the crypts, some of which are as big as small cottages, filling the necropolis with life. On a hill above Cairo we visited the Azhar Mosque, built more than a thousand years ago, its minarets rising above the polluted air of the teeming city below.

Dominating the city are five wonders of the world, one natural, the Nile, and the other four man-made, the three great pyramids and the Sphinx. The pyramids are so large that most of the old city of Cairo was built from their stone facing, yet still they stand, appearing unaffected by the most extensive and egregious act of vandalism ever committed. The pyramids extend back five millennia. They are so massive, and the climate so arid, that they are almost certainly the most enduring structures that will ever be built by our species on this planet. The feeling of great age associated with the pyramids was palpable. The Egyptians say that "man fears Time, but Time fears the pyramids."

We sat on the Great Pyramid of Cheops, or Khufu, and heard, then saw, tomb bats *(Taphozous perforatus)* taking refuge amidst the huge stone blocks. Tomb bats have been living in pyramids since the first ones were built. They undoubtedly entered Khufu's tomb at about the same time as the pharaoh Khufu himself if not before. Now the ancient Egyptian civilizations of pre-biblical days are gone, their voices stilled thousands of years ago. Only the pyramids remain, and the tomb bats. Each night the bats leave their tombs to forage on insects that fly above the silent Nile, just as they did when the pharaoh was overseeing the construction of his tomb. Each morning they return to rest during the hot Sahara days in the eternal darkness of the pyramids.

I rented a horse and rode 20 miles south across the desert past the great pyramids to the first pyramid ever built—the step pyramid of King Zoser at Saqqara. Riding past the hordes of beggars crowding the narrow roadways near the edge of the city, I had a hint of what it must have been like for the early conquistadors, crusaders, and conquerors to ride through masses of people who did not have horses. Towering above the street-dwelling poor, and able to see far across the dense crowds, the horse cleared the way and the people fell back. I experienced a most unexpected feeling of power—not a welcome feeling, but one that was enlightening.

The Arabian horse was hard to handle, but eventually I was able to make my way through the city and race across the dunes toward Saqqara. The sound of the booming hooves as I rode over the dunes was one that I shall not forget. From Saqqara, I could see the bent pyramid and, far in the distance, the red pyramid, structures rising from the arid landscape that proclaimed its antiquity. No other structures were visible as I looked into the distance over a desert landscape that had probably changed little since the time of Cleopatra.

On a Sunday evening before heading out to the desert Duane and I set bat nets in a mango orchard along the Nile. The night was silken, with the scent or ripening mangoes filling the air and the ancient river flowing placidly by only a few yards away. Soon we heard the wingbeats of bats and collected an Egyptian fruit bat (*Rousettus aegyptiacus*), my first megachiropteran.

There are two great divisions of bats, the microchiroptera, or microbats— which are generally small, are the most numerous, and occur worldwide— and the megachiroptera, or megabats, which are generally large and live only in the Old World. Some megabats are called flying foxes. Unlike microbats, most of the Old World giant fruit bats do not use echolocation, the sonar pulses that permit microbats to fly through caves or dense foliage at night. Using sonar, microbats can feed on flying insects in absolute darkness by bouncing ultrasonic pulses they emit from their mouths off objects in their path and analyzing the reflected sounds. Lacking echolocation to assist in night flight, and feeding mainly on fruits, some species of giant bats fly in full daylight. The Egyptian fruit bat, however, was one of the species of megabats that did use echolocation, and as darkness fell the bats flew unerringly through the mango trees.

When we left Cairo to search for study sites for the long-term research project, we traveled first to Wadi Natroun, which is 50 miles west northwest of Cairo. The Sahara in this region is as sparse as any desert I had so far encountered. In some places there were few plants and those were only a foot or so high. In other places there were no plants for miles. Pebble desert (called hamada in the Old World), where small pebbles form a crust over the desert soil, was common, as were dunes.

The most notable thing about this part of the Sahara is the golden color of the sand and of the desert in general. In some deserts, the soil is white. In others it is pale brown. In still others the soil may be red or even black. In the Sahara, however, the soil has a golden tinge, lending great beauty to the landforms. Perhaps the crystalline air and intense ultraviolet rays of the sun in the open desert accentuate the color.

In the evening we set out 130 traps to find out what small mammals lived in the area. We were cautious, for both sidewinders (*Cerastes cerastes*) and Egyptian cobras (*Naja haje*) are present. Unlike most poisonous snakes in North America, neither of these snakes rattles. The sidewinder is small and not especially deadly, but a bite from the cobra would almost surely kill a human quickly if he could not soon get to a hospital that stocked antivenin. In the morning we had caught 26 rodents, including four types of gerbils: 15 pallid gerbils (*Gerbillus perpallidus*), 2 Anderson's gerbils (*Gerbillus andersoni*), 3 lesser gerbils (*Gerbillus gerbillus*), and 1 gerbil known charmingly as the pleasant gerbil (*Gerbillus amoenus*).

We had also captured 5 lesser jerboas (*Jaculus jaculus*), bipedal dipodids that are closely related to the three-toed jerboa, the species I had captured in Iran. The lesser jerboa is almost a perfect reflection of the kangaroo rats of North America—long hind legs, small forelegs, a tail with a white tuft on the

A lesser jerboa (*Jaculus jaculus*), a Saharan analogue of North American kangaroo rats. (Photo by M. Andera; Mammal Slide Library, American Society of Mammalogists.)

end, similar body size, and obvious bipedality. It includes seeds in its diet, but also eats a great deal of green vegetation and roots. The jerboa, living in an extremely arid habitat, is known to travel long distances in search of food. One animal was followed on a trek of more than 8 miles in a single night, a remarkable distance for a small rodent (most rodents of similar body size travel only a few hundred yards in a night).

A 20 percent capture success and so many species taken in one area showed that the Sahara was no Monte Desert. There were many species and several were abundant. Most gerbils were seed specialists, like the kangaroo rats and pocket mice of North America. This fact was well known, and I was not encountering any species that were new to science or making observations that had not already been made. What I was doing was bringing a global view to the desert rodents of the Sahara, comparing them in many ways with the rodents of other deserts. After obtaining first-hand experience with them, I would be able to extend my morphometric studies to the animals of the Sahara with confidence in the information revealed through the analyses since I would have learned about the animals in the field.

Although there was information available on the natural history of some of the species, there had been very few field ecology studies conducted on Saharan small mammals. It was a desert that cried out for extended research. Indeed, other than in the deserts of the United States and, in recent years, Australia and Israel, the ecology of small desert mammals has mostly been ignored. Few biologists had done long-term ecological research on small mammals in the Sahara and I hoped to be among the first.

When we checked our traps we occasionally saw helicopter gunships fly by, or were startled by the sudden roar of jet fighters. There were two airbases located nearby and both were on a war footing. Three months earlier, only 6 miles from our camp (less than a minute away by jet), a North American oilman had strayed off course in his Jeep and been killed by gunfire from a helicopter gunship. We were told that we had to stay clear of power-line poles, which had land mines placed all around them, and avoid driving across open desert, since any car that did so could become a target for one of the jets or helicopters that patrolled the area. I made two notes in my field book: "Have to check with military because jets will wipe you out. Mines around towers . . . by aiming between them one should get through all right." We were near the small town of Bir Hooker, which is the main town in the Wadi. I wrote, "The town is not much and makes Exu look like Rio by comparison."

We continued to explore the area to find appropriate study sites. Plants included small acacias and shrubs of the goosefoot family, that widespread family that includes saltbush. One Sunday we visited the Monastery Deri el Bishoi, the monastery where Saint Bishoi began his priestly life. Bishoi was an apostle of Saint Anthony, the founder of the Coptic Christians in about the third century C.E. The monks at Bishoi's monastery were friendly and, in principle, were not averse to tents being pitched either just outside or even within the monastery walls. We inquired about renting rooms at the monastery some time in the future, thinking it would serve as a good field base when we were away from the main camp. An added advantage was that the ancient monastery was unlikely to be bombed or strafed by military aircraft. One of the monks took me to see Saint Bishoi's coffin, which was in a small chapel. The coffin was tiny. Bishoi could have been a jockey.

"Bishoi believed in suffering," said the monk. "He denied himself sleep throughout his life. He never complained."

As one who has always found sleep elusive, I agreed that giving up sleep was certainly a great sacrifice, but then said, "Everyone has to sleep."

"He had long hair," said the monk, "and at night he would use a rope to tie his hair to the rafters so that if he fell asleep, he would be startled awake as he began to fall and the hair would suddenly be pulled tight."

"That would do it," I said, "but of course, he would have had to take a nap at some point."

"No, he never slept, not ever."

We had reached an impasse. I figured Bishoi had become the master of forty winks, probably surpassing Thomas Edison or your pet cat in his ability to snatch and conceal a quick nap, but there was no sense in going into it any further. I prepared to leave the chapel. The monk was not finished, however. He said:

"His body is uncorrupted."

I paused. Now this was starting to get interesting.

"Really? How do you know?" I asked.

"We all know this."

"So in that little coffin is Bishoi looking as fresh as the day he died more than fifteen hundred years ago?"

"Yes. This we know."

The coffin had a simple latch on it.

"May I look at him?" I asked.

The monk was taken aback.

"Of course not," he said.

"Have you ever seen him?" I asked.

"No."

"Would you like to see the saint?"

"No. I do not need to see him."

"But would you like to?"

"No."

"Do you know anyone who has seen him?"

"No."

"How long has it been since he was seen by anyone?"

"More than a thousand years."

"I would really like to take a peek at him. Would you leave the room and allow me to crack the lid of the coffin just a tiny bit? I will not tell you what I find inside."

He was aghast.

"No, we must leave now."

"Do you ever doubt that he is uncorrupted?"

"Never."

I believed him. If I were a monk, however, I would have peeked in the coffin shortly after arriving. This indiscretion would have resulted in my discovering bits of jerky-like skin and some brittle bones of Saint Bishoi, which is surely all that was left of the long-suffering monk after so many centuries. My faith would have been sorely tested then.

Perhaps daring to look and then dealing with what one finds is a greater test of faith than merely believing and refusing to look. If that is so then a scientist may have greater faith than a monk. Perhaps going on with life when one knows there are no miracles, only mysteries that are as yet unexplained, is the truly difficult road. It is far more difficult to search for the fact of the thing than to ascribe facts to a higher, unknowable authority. Viewed in this way the apparently difficult life of the monk was easier than my own.

Moreover, in my experience people who do not sleep *always* complain. So if Bishoi did not grumble about having his hair tied to the rafters, then he must have been a saint. A more likely scenario is that those who had to put up with his constant complaining about not sleeping were the true saints. The monks believe that he did not sleep and did not complain. If he could do that and all of his other good works as well, he was surely a saint. If he was a saint then he did not decompose within the tiny coffin in the desert chapel. The logic is impeccable.

The monk who guided me through the monastery had no need to look inside the coffin. The very act of looking would be evidence of a loss of faith. It could prove nothing except that he had followed the wrong calling. This is also logical. Indeed, it is a chain of logic that is irrefutable for the religious, but unacceptable, and almost incomprehensible, for the scientist. Desert monks and desert biologists deal in two separate realities, and the differences are especially pronounced when one is confronting "facts" that oppose biology.

We continued to sample the small mammal fauna of Wadi Natroun. I caught a most unusual rodent, the fat-tailed gerbil (*Pachyuromys duprasi*). This small desert mouse, weighing in at 1.5 ounces, lives in bare pebble desert. To understand how bare the habitat is, imagine that you have had all of the concrete removed from your driveway, so that only the compacted dirt remains. Then scatter a few wheelbarrows full of small gravel over the dirt. Finally, imagine 10 square miles of this surface with no plants whatsoever sprouting from it. You have just recreated the habitat of the fat-tailed gerbil.

The day before I had thought, "Now here is a spot where I would never set traps." That is the problem with ignorance. You do not know what you are doing, but you have a good reason for doing the wrong thing. I told myself, as I often do when I am in a new desert, "I know nothing. How can I be certain that some mammal has not managed to find a way to live in this apparently lifeless gravel plain?" So I dutifully set a few traps in the seeming wasteland. Had I walked on, I would have missed the fat-tailed gerbil.

The fat-tailed gerbil is a lovely pale blonde rodent with a broad head, a pure white belly, big black eyes, and a pointed snout. The broad head is the result of the skull having enormously inflated tympanic bullae. They are even more inflated than those of kangaroo rats. Indeed, they are more like those of the plains vizcacha rat of the Monte Desert. In some areas, fat-tailed gerbils live in sandy zones where plants such as sagebrush (*Artemisia*) are present, although very sparsely distributed. The animals are known to eat fruits from desert plants, seeds, and some green vegetation. In captivity they have shown a fondness for crickets. Some have been known to travel up to

The habitat of the fat-tailed gerbil (*Pachyuromys duprasi*), Wadi Natroun, Egypt.
(Photo by M. A. Mares.)

1.6 miles across their bare habitat to collect fruits and seeds, quite a trek for such a tiny mouse. Like all gerbils, these are specialized for life in the most arid parts of the desert, and are similar in many ways to the North American pocket mice.

After completing fieldwork in Wadi Natroun we traveled to a spot west of the city of Alexandria. The Mediterranean was the color of a blue diamond and the desert's color had also changed. Gone were the pebble deserts and golden sands. Near the Mediterranean the sand became white, its color accentuated by the cobalt sea. The sparse shrubs were gone, and in their place grew numerous saltbushes. Salt desert ran down to the sea.

I walked through part of the desert and came upon a Bedouin village, an unusual occurrence, for Bedouins are perhaps the quintessential nomads. In recent decades many North African governments have been trying to get the Bedouins to give up their nomadic ways, but with only limited success. The Bedouins I visited were cultivating figs along a narrow strip of land between the saltbush flats and the sea. I spoke with a lovely young Bedouin girl who offered me a fig and told me the history of her people as I reveled in the beauty of the Sahara's edge as it dropped into the Mediterranean Sea.

The Egyptian coast of the Mediterranean is fascinating. As we worked our way through the desert we came upon a Roman fortress built centuries before the time of Christ. The Greeks had traveled through the region, too, and the ruins that mark their passage now dot the desert. We came upon a salt flat, very much like salt flats in other deserts. The saltbushes grew on small hillocks of sand, just as they did in the Great Basin of North America and the Monte of Argentina. Along the edge of the salt plain were the mounds of dirt that signaled the presence of the mole rat (*Nannospalax ehrenbergi*), animals so unusual that they are usually placed in their own family, Spalacidae (although some authorities think that they belong to the family Muridae, the Old World rats and mice).

Mole rats are burrowing rodents like the gophers of North America or the tucu-tucos of South America, but they are even more specialized for life underground. For one thing, they are blind, their eyes covered with fur. Their large upper incisors project forward and, along with their broad front feet, are used in digging. They have no external ears or tail, and are covered in dense, short fur. In essence, they are a tubular mass of tissue that is specialized to dig tunnels more than 100 feet long through the desert soil and feed on tubers, roots, and bulbs, which they store underground in special chambers. They also dig chambers that they use as latrines, and one that is used as the nest. Like almost all rodents that are highly specialized for life underground, they are solitary and have a nasty temperament. They can reach high densities for a burrowing desert rodent, and their numbers in the deserts of Israel were estimated at about 2 million per 15,000 square kilometers.

The salt content of the soil increased and the vegetation became more halophytic as we moved further into the edges of the salt flat and away from the sandy desert scrub. Saltbushes became the dominant plants. Under the saltbushes were the burrows of the fat sand rat *(Psammomys obesus)*, a relative of gerbils that occurs only in areas that support salt-loving plants. We had no time to capture the animals, but it was clear that these woodrat-sized rodents (5 ounces) that looked much like the plains vizcacha rats of the Monte were also specialized to feed on salt-filled plants. Like squirrels, the sand rats are diurnal, but spend time in the burrow during the heat of the day. The animals do not live in colonies, but are aggressively antisocial and solitary.

I was never able to study sand rats, but physiologists had studied the species in Algeria. They found that the animals were highly specialized for feeding on the salt-filled leaves of chenopods, the same types of plants eaten by the chisel-toothed kangaroo rat of the Great Basin Desert. Their kidneys were also specialized to rid the body of salts. A later study showed that the lower incisors of the fat sand rat had the same chisel shape as those of the salt-eating kangaroo rat and that the animals used them to strip away the salty plant tissues in the same way the North American species did.

Until the 1990s, when my colleagues and I discovered the salt-eating behavior of the plains vizcacha rat of the Argentine desert, the chisel-toothed kangaroo rat and the fat sand rat were the only known salt specialists among desert rodents. They had converged on each other in adaptations of the teeth and kidneys. Of course, the vizcacha rat had gone a step further and developed brushlike salt strippers out of stiffened hairs that function like an extra pair of teeth. No other mammal had taken that unusual evolutionary step.

In the coastal desert of northern Egypt my colleagues and I found that the four-toed jerboa *(Allactaga tetradactyla)* and the greater Egyptian jerboa *(Jaculus orientalis)* occurred together. Thus most of the major types of desert rodents were present in Egypt's Western Desert. There were bipedal species, much like kangaroo rats (but feeding mainly on green vegetation and underground plant parts). There were small, nocturnal seed specialists (most of the gerbils, which were ecologically similar to pocket mice). There was a diurnal leaf eater specialized on saltbushes (the fat sand rat, which filled a niche similar to the chisel-toothed kangaroo rat or the plains vizcacha rat of the Monte Desert). A species of burrowing rodent (the mole rat) was present that ate roots and tubers, like the gophers or tucu-tucos of New World deserts. Several other mammals, such as shrews and hedgehogs, also lived in the desert and fed on insects and other invertebrates. The vast Sahara was much more like the deserts of North America in its diversity and abundance of species than arid North America was like the Argentine Monte. The Sahara also supported many more species of desert rodents than did the Iranian Desert.

We worked to develop a grant proposal to study all aspects of small mammal ecology in the Sahara Desert. Our plan was large and ambitious and would have involved many Egyptians and Americans. We were filled with enthusiasm and optimism, for no one had taken on the Sahara in quite this manner. Our proposal was submitted to the agency of the U.S. government that provided funds for research in countries that had debts pending from World War II. The interest on these debts could be used to support research in the country. The proposal was approved. We were ready to begin. It all seemed too easy—and it was.

The war in the Middle East that was under way even as we visited El Alamein had escalated shortly after our departure from Egypt in 1977. Henry Kissinger had begun his shuttle diplomacy, flying repeatedly from Egypt to Israel and the United States, beginning in 1974 under President Nixon and continuing under President Ford. President Carter committed the United States to heavy involvement in the Middle East peace mission. Jimmy Carter, Egyptian President Anwar Sadat, and Prime Minister Menachem Begin of Israel worked hard to obtain a peace accord in 1979. It was a remarkable display of diplomacy aimed at averting greater conflict in the war-torn region. Unfortunately for us, the money that covered the enormous cost of the diplomatic effort was the same money that had been destined to support our research. When peace came, our project died as surely and as finally as King Khufu. To this day no one has carried out a project of similar scope.

My colleagues' and my efforts to develop a long-term project in the Old World deserts were frustrated at every turn. Egypt no longer had funds to support research on desert mammals. Our Iranian project died with the capture of the U.S. Embassy by followers of the Ayatollah. We turned our attention to Afghanistan, another country with extensive deserts, but the Soviet

Union invaded at just about the time the Egyptian-Israeli peace accord was signed. We also considered Pakistan, another country with an unstudied desert fauna, but the government of President Zulfikar Bhutto was taken over in a military coup in 1978 and the generals who imposed martial law hanged him.

I joked that I was afraid to suggest research in the Soviet Union or China, for that would surely lead to World War III. It seemed that the geopolitical forces of the planet were aligned to thwart our efforts at comparative desert research. Because of political instability, lack of funding, or my own changing professional responsibilities, I have still not been able to develop a long-term ecological research project on mammals in an Old World desert. To this day, no modern inclusive study on small mammal ecology has been conducted in Egypt, Iran, Afghanistan, or Pakistan.

Naming the Anonymous

Taxonomy . . . is a craft and a body of knowledge that builds in the head of a biologist only through years of monkish labor. The taxonomist enjoys the status of mechanic and engineer among biologists. He knows that without the expert knowledge accumulated through his brand of specialized study, much of biological research would soon come to a halt . . . If a biologist does not have the name of a species, he is lost . . . A skilled taxonomist is not just a museum labeler. He is a world authority, often *the* world authority . . . steward and spokesman for a hundred, or a thousand, species.

Edward O. Wilson, *Naturalist*, 1994

I was able to return to Argentina for extended fieldwork several times in the late 1980s and 1990s. I wanted to expand my understanding of the fauna of the country as a whole and explore hidden corners of the Monte and Andean regions that I thought could harbor new species. For many years *Andalgalomys olrogi*, the mouse genus I had discovered in 1972, was known only from the Amanao River in Catamarca Province, the type locality. In November 1987 my colleagues and I collected the first *Andalgalomys* captured outside of the valley of Pipanaco near the town of Chumbicha, also in Catamarca Province, but further east across an imposing mountain range. As I wrote in my field notes: "The mouse caught yesterday is some type of *Andalgalomys*—maybe *olrogi*, maybe not. It'll bear some looking into when I get back."

Map of Argentina showing some sites I visited on research trips. (Map by Cartesia MapArt, modified by Patrick Fisher.)

When in 1990 we were finally able to capture additional animals in San Luis Province, to the south of the original locality, we found that we had discovered a second Argentine species of *Andalgalomys*. My colleague Janet Braun and I named the new species *Andalgalomys roigi* for our friend and colleague Virgilio Roig, who had studied mammals in the Monte Desert for almost a half century and had dedicated his life to scientific research and conservation.

We surveyed the mammals of all of Argentina in the early 1990s with the support of a grant from the National Science Foundation. Funds were granted for discovering new mammals, for learning new facts about poorly known species, and for increasing the numbers of museum specimens of Argentine mammals. Museum specimens were then increasingly being recognized as forming an irreplaceable database of information about species during a period when they were disappearing throughout the world because of habitat destruction. Moreover, good systematic information on mammals, especially rodents, was being recognized as vital to medical research on newly emerging viral diseases and other maladies.

As our work progressed we returned with increasing frequency to the Monte Desert and the bordering Andean and pre-Andean habitats, for I knew that we still had a lot to learn about the mammals of that interesting region. My students and colleagues carried out much of the work, but I made time for at least one field trip every year. It was difficult to get away for extended periods because I was immersed in the process of developing, funding, and building a new natural history museum for the University of Oklahoma, an endlessly demanding job that consumed my energies the way a black hole absorbs light.

I vowed that I would not let my research activities suffer because of administrative duties, but it was not an easy promise to keep. Each time the field crews left for Argentina, I moped for days. I did not want them going to all those neat places without me. I wanted to be there when the habitats were selected, the traps set, the animals captured. I was missing out on a great adventure. (Curiously, building a new museum turned out to be quite an adventure, too, but that is another story.) Ironically, for most of my life I had lacked adequate financial support to do the kind of field research that I wanted to do. My crews and I usually had to scrape by to get into the field, sometimes eating the same species we were collecting because there was no money for food. Now, when I finally had the funds to do the research in the proper manner, I was forced to spend most of my time carrying out my administrative responsibilities. For many months every year, I yearned to be in the desert as my field crews worked without me.

A lot of planning went into each trip even before the crews left the United States. Using my knowledge of the country's habitats and mammals and my gut feelings for where new mammals might be found, I marked areas on maps for the crews to visit. I had to leave the final decisions as to where to camp and trap up to them, of course, but I told them how they should get to the areas, how long they should trap in each locality, and the kinds of animals they might expect to capture. I was reduced to the role of mother hen. One place I wanted them to go was an area containing several isolated salt flats in San Luis Province in central Argentina. I tried to send them to areas that I had never visited, but I had covered so much of Argentina in my travels over the decades that I had some familiarity with most of the country's habitats. I sent the field crews to a variety of poorly sampled and isolated habitats that looked promising.

On the first field trip to one of these areas the crew collected a small mammal that they recognized as being different from all the others. When I arrived to lead the next field trip, I examined the collection and concluded that they had collected a new genus and species. Over the next couple of years we were able to collect several more individuals and finally to determine its preferred habitat. The animal had a long tail and large hind feet, and seemed a likely candidate for the hypothetical bipedal Monte Desert mouse. Could it be my long-sought South American equivalent of the bipedal North American kangaroo mouse? Eventually we captured one alive and took it to a salt flat, where we let it run free. The animal did not hop. Instead, it spied the only tree growing on the salt flat and raced toward it, quickly climbing into the tree with the practiced agility of an arboreal mammal. Its behavior suggested that trees were no strangers to this little mouse.

We were able to pinpoint this species' preferred habitat simply by noting where the animals were most abundant, which was in a relictual forest of Chacoan trees near the vast salt flats of west-central Argentina—an isolated habitat separated long ago from a larger habitat that is no longer close

The delicate salt mouse *(Salinomys delicatus)* from San Juan Province, Argentina. (Photo by M. A. Mares.)

by. Janet Braun and I named the genus and species *Salinomys delicatus*. *Salinomys* refers to the "mouse of the salt flats," and *delicatus* to its delicate body form. It is a beautiful gray mouse, and probably spends most of its time in the isolated scrub forests that were left behind as the Chacoan forest withdrew during the increasingly arid Pleistocene. During wetter periods the Chacoan forest extended into much of today's Monte Desert. The forests were extensive, especially near the salt flats, where water still accumulates today. But as the climate changed the forests retreated eastward, leaving scattered remnants in the wetter microhabitats of the Monte Desert.

Today, the isolated patches of forest harboring this unusual rodent are disappearing because they are being cut down for firewood. It is only a matter of time until the forest where the type specimen was collected disappears, taking the delicate little mouse of the salt flats with it. It is no longer anonymous, but its biology remains unknown.

Our investigations of the mammals of Argentina continued throughout the 1990s. In 1995 we climbed above the city of Mendoza into the isolated valley of Uspallata. The area is especially lovely, with the Mendoza River carrying snowmelt to the lower valleys of the Monte. Uspallata is at 7,000 feet of elevation and is surrounded by arid but breath-taking mountains. The tallest peak in the New World, Aconcagua (22,000 feet high), is part of the mountain chain that rims the valley. The valley itself is arid, with Monte vegetation in the basin and cacti and sparse shrubs on the hillsides. The underlying minerals have been exposed after being folded and refolded upon themselves in massive geological shifts, and the steep slopes display an array of colors. The valley and the surrounding mountains were used to represent Tibet in the movie *Seven Years in Tibet*. Indeed, the Andean highlands of Argentina look very much like the Tibetan Plateau. Both are high deserts dotted with sparse, low shrubs and surrounded by massive mountains.

We learned of an unusual rodent from a botanist we met who had taken a picture of it while riding his mule in the area. He took the photo in an extremely arid valley, the Quebrada de la Vena, that joins the valley of Uspallata. As he examined the vegetation thereabouts, a small rodent scurried out of the rocks. We could tell from the photo that the animal the man had seen was a chinchilla rat *(Abrocoma)*, the only living genus in the family Abrocomidae. Among South America's rodents, very few species belong to a genus that is so unusual that it is the only genus in a family, but the chinchilla rat is one of the exceptions.

When we entered the Quebrada de la Vena, we were struck by how the already dry habitat of Uspallata became much more arid in this narrow valley surrounded by multicolored slopes supporting little vegetation. On the hillsides were many massive rock faces with deep fissures that had formed as the rock had cracked. In some of these fissures we saw thick, reddish-black piles of fossilized excrement extruding from the rocks. Some of these middens were more than 3 feet high and their surfaces felt like hard plastic. They had clearly been accumulating for centuries.

In the rocks above the valley another unusual rodent watched us work. Periodically, we saw mountain vizcachas sitting on boulders far above where we were trapping. *Lagidium viscacia* is a member of the chinchilla family, Chinchillidae (and no relation to the chinchilla rats). It could almost be called the "lucky chinchilla," because this large, rodent—weighing 7 pounds—molts its beautifully soft fur throughout the year. Its smaller cousin, the chinchilla *(Chinchilla laniger)*, which also lived in these mountains many years ago, was almost driven to extinction by fur trappers who collected the animals for their uniquely rich and soft pelt. Chinchillas did not molt throughout the year and their pearlescent pelts were coveted by furriers. Fortunately the trappers did not trap all of the chinchillas. A few remained in small populations in extremely remote areas. Today these populations are being rediscovered and protected in the Andes of Argentina and Chile.

A mountain vizcacha looks like a rabbit with a long tail that sports a prominent black or dark reddish flag formed by a thick ridge of stiff hairs. These rodents blend in with the surface of the boulders, where they spend a good deal of time sunning and watching for predators. Eagles, hawks, foxes, and wild cats are common in the area and the vizcachas are a major food source for all. Mountain vizcachas are colonial, living in grottoes and burrows in and under large boulders. In some respects they resemble North American marmots, for they are usually found in groups and are often seen foraging on grasses and shrubs growing at the base of the steep mountain slopes. They are active only during the day, emerging to sun themselves after the sun has climbed high in the sky. When the cold shadows of evening fall across the boulders they retreat into their burrows to escape the long, cold night.

We set up camp in the isolated and uninhabited valley and tried to collect chinchilla rats. It was a difficult place to trap, for the hillsides were extremely steep and were composed of loose rock and shale that made climbing treacherous. We had to scramble up the slopes on all fours, and often the shale gave way, making us slide back several yards. Then, like Sisyphus, we had to regain the lost ground. The animals could hardly have chosen a more inaccessible place to live. We were able to reach only a few dens, those in rocks that we could reach after a steep, dangerous climb up the slopes. Most of the dens were in cliff faces high above the valley and could be reached only by using rock-climbing apparatus, which we did not have and would not have known how to use if we had had it.

We did not know if the *Abrocoma* were primarily diurnal or nocturnal. Diurnal animals are usually harder to trap than nocturnal ones, since the traps can hardly surprise them. We knew that for at least part of the year they were diurnal because our photographer had snapped the photo in full daylight. We set various types of cages and rat traps for the animals, but for several days had no luck in capturing one, or even in seeing one. Finally, one morning one of the traps held an *Abrocoma*. After examining and photographing the animal alive, we sacrificed the specimen and karyotyped it, taking a sample of its chromosomes for genetic analyses. We also prepared various tissues for use by other researchers, for whom living tissues are a treasure trove of information. It was finally time to prepare the animal as a museum specimen, but before I did, I wanted to photograph it again. As I was handling it, I noticed that the animal had a most unusual characteristic. It had two sets of false teeth!

All rodents have two upper and two lower incisors, which are their gnawing teeth, perhaps best exemplified by the large incisors of beavers. Rabbits, which look like rodents, appear to have two pairs of incisors, but actually have two pairs of upper incisors and one pair of lower incisors. Situated behind the large upper incisors is a tiny second pair of incisors. This is one reason rabbits are placed in a separate order from rodents.

This chinchilla rat had two extra sets of upper incisors, both in the midline of the skull—a larger pair located toward the front of the mouth and a smaller pair behind it. The "teeth" were hard, white, and shiny, like real teeth, but they did not attach to the bones of the skull. They grew out of the soft tissues of the mouth. It was an extremely odd finding and had never before been described for a mammal. The animal also had deep, hardened ridges on the upper palate, and the tongue had a hardened grinding pad on its dorsal surface. The palatal grinding surface, along with the hardened tongue pad, apparently functioned like a mortar and pestle, efficiently pulverizing plant matter.

The Uspallata chinchilla rat *(Abrocoma uspallata)*, a new species from Mendoza Province, Argentina. (Photo by M. A. Mares.)

Clearly, these are adaptations for a specialized diet. Most of the plants growing around the boulders were creosote bushes or cacti. The middens suggested that creosote bush was the primary food, which would help explain the resilience of the middens to decomposition. Creosote, after all, is a remarkable preservative. The extra "incisors" serve a function we cannot deduce without obtaining additional specimens and observing them alive.

We were unable to capture more individuals, although we determined by censusing middens that the rats are common in the area. Apparently, they are very good at avoiding traps. Unfortunately, we do not know if all chinchilla rats have similar false teeth. We do not know what the animals eat, whether they are diurnal or nocturnal, when they reproduce, or how they survive the long, cold Andean winter. We do not know if they live in colonies (which one might surmise, given the extent of the middens). Do they hibernate? Are they long-lived? Do the offspring or relatives inherit the rocky crevices so that the middens reflect the cumulative results of lineages over time? We know almost nothing about one of the oddest animals I have ever encountered. Finding the answers will surely provide startling insights into the adaptability of mammals to the challenging conditions of aridity. The species remains in that isolated valley, harboring its mysteries for the next field biologist who is willing to spend more time in one of the most picturesque valleys of the Monte.

A few years after that trip I was able to go to London to examine the type specimens of the various species of chinchilla rats from Argentina and Chile that had been named by Oldfield Thomas. I found that our specimen was a new species. We named it for the beautiful valley of Uspallata, *Abrocoma uspallata*, a valley that looks like Tibet, but whose fauna is purely Andean.

When I first went to Argentina I considered myself an ecologist, but since I regularly encountered animals that I had difficulty identifying, it quickly became clear to me that without a strong foundation in taxonomy most ecological research could not proceed. Consider the situation of the gerbil mouse, *Eligmodontia*, that was described earlier. If only a single lowland species existed, as had been thought, it would have been simple to study the animals. Indeed, you could study them anywhere, whether in the altiplano of Argentina or along the coast of Patagonia, for the behavior, reproductive biology, or physiological adaptations that you examined would characterize a single species—any differences found would simply be variations within a single species.

But what if the prevailing taxonomy was incorrect and the genus included more than one species? What if some were high-altitude specialists and others lowland desert, sand dune specialists (which is what we were discovering)? If animals in one region had a totally different reproductive pattern or behavior, how could you interpret the data? What if two species of *Eligmodontia* occurred together in an area, but you thought that they were the same species—something we also found with gerbil mice in the Monte Desert? How could you learn that species A inhabited sandy areas, whereas species B lived in soils that were more compacted? What if species A carried a pathogenic virus that was transmissible to humans, but species B did not?

How could a biologist predict where the disease would spread? How could anyone develop a control program?

The genus *Eligmodontia* contains at least eight species. Even the type species is open to question. The great French zoologist Georges Cuvier, the father of comparative anatomy and paleontology, named *Eligmodontia typus* in 1837, but the type specimen vanished some time during the next 150 years. Moreover, the locality for the type specimen was listed as "Corrientes," a province which consists largely of swampland and in which this little desert animal does not occur. By listing Corrientes as the locality, the collector probably meant that the specimen was obtained somewhere in Argentina. If there were only one species in Argentina the name would be no problem; but there are many. The same year Cuvier named the species, the British anatomist Alfred Waterhouse named a specimen collected by Darwin on the voyage of the *Beagle*. He called it *Mus elegans*. It was *Eligmodontia*, but Waterhouse failed to recognize it as a distinct genus. The species name used by Waterhouse was perhaps more fitting for this elegant little mouse than Cuvier's *typus*, which merely meant that it was the type specimen of the genus *Eligmodontia*.

Unlike Cuvier's lost specimen, Waterhouse's specimen still exists; it is in the British Museum of Natural History. I have studied the specimen collected by Darwin almost two centuries ago. As Darwin wrote: "This little animal does not appear to agree exactly with any of the subgenera of Cuvier. It was caught Octob. 3ᵈ at Monte Hermaso in B. Blanca. In bringing at night a bush for fire wood, it ran out with its tail singed." Under the rules by which species are named, the type specimen of the genus may become the one collected by Darwin, since it cannot be determined which species Cuvier actually named or where it was collected. Those specimens that are today known as *Eligmodontia typus* will be known as *Eligmodontia elegans*. Taxonomy has rigid rules that are designed to impose some sense of order on the bewildering diversity of life. If the name is changed officially—and the change is accepted by the group that regulates taxonomic nomenclature—the little gerbil mouse will no longer simply be "typical." It will become "elegant," perfectly fitting the beautiful desert mouse captured by Charles Darwin on one of history's great voyages that changed the world.

If you do not know the taxonomy and systematics of the organisms you study—if you cannot identify them correctly and understand how they are related—then you cannot study them in any meaningful manner. Without a proper taxonomy scientific studies become chaotic. Without systematic studies that show how organisms evolved and are related, you cannot grasp how they came to be, their patterns of colonization of an area, or their evolutionary history. And if you do not understand the evolution of a group of organisms, you cannot decipher their interactions through time with other organisms such as plants, parasites, and viruses.

Not to understand how a species differs from a closely related species is, therefore, to be ignorant at a most fundamental and profound level about all aspects of the biology of the species, its populations, the community of interacting species, and the ecological system in which they function. If you did not know that the chisel-toothed kangaroo rat was a distinct species, the interesting story of the saltbush and the rodent could never have been understood. The fewer such stories you can elucidate, the greater your ignorance of nature. If you do not know the taxonomy, you know nothing.

Unfortunately, since most scientists today are trained in and work in developed nations, they no longer learn to identify animals correctly, for most of the taxonomic problems in developed countries have already been clarified by the field biologists and taxonomists of yesterday. If I were to initiate a study of kangaroo rats, for example, I would find it simple to identify my study animals because the taxonomy and systematics of the twenty-one North American species have already been clarified. I could obtain a field guide to the mammals in any bookstore in the country. Kangaroo rats and most other small mammals are now easy to identify. So if a student wants to study chisel-toothed kangaroo rats, he or she could easily distinguish them from other kangaroo rats by comparing them with the few species already known to occur in the area. As long as the student can identify the study animal, the research can be carried out without any additional training in how to identify species.

In North America this is the rule. Learn a few species, and your ecological study can proceed. In those less-studied parts of the world, however, this is not the case. If you dropped the vast majority of North American mammal ecologists into South America, Africa, or Asia, they would probably be unable to identify *any* of the mammals they encountered. Without a specialist who knows the group, they would be adrift in a sea of unknown species, unable to conduct ecological research, or they would misidentify the study animals and proceed with a flawed study, polluting the scientific literature with nonsense. I have seen both things happen.

If investigators conduct ecological research on species they cannot identify, the study is doomed. Unfortunately, young biologists no longer learn to identify species. They do not learn about taxonomy or systematics. Most do not take the classic courses that provide a strong background in species identification, natural history, field methods, and zoogeography, courses such as mammalogy. They concentrate on synthetic courses, or on working with only a few species, becoming specialists before they are generalists. Most of them work in developed countries with well-studied mammals, so they can proceed with their research with only minimal problems. They are sublimely ignorant of the diversity and complexity of nature, and they limit their research to those species that were studied long ago by zoologists who outlined the evolutionary relationships of the species.

This is especially frustrating to me because taxonomy and systematics are considered less important, and even less scientific, than other fields of biology by many modern ecologists and conservation biologists. They often do not understand that their research is only possible because someone has delineated the limits and the evolutionary relationships of the species they study. This trend began in the 1960s when ecology underwent a metamorphosis into a more experimental, less theoretical, science.

I am not alone in such views. In a fascinating paper in *Tropical Forest Remnants*, Francis H. J. Crome commands:

> Recognize the importance of high-quality natural history studies. By natural history, I mean the sort of descriptive studies characterized by deep thinking and the application of logic that were common in the nineteenth century. Such studies are rarely funded today, yet they would provide the kind of background on species' behavior and distributions that would make quantitative research

much better, and would also provide at least some of the information necessary for decision making . . . Be suspicious of all but the most obvious generalities . . . We must accept that ecological data is often very rough, and we can't get much out of it.

Our training and our experience with nature influence all of us. If we know only one fauna or one group of organisms, how much do we really know of nature? Fra Mauro, a monk who was cartographer to the Court of Venice in the sixteenth century, was apprehensive when he prepared to draw the latest official map of the world. On his shoulders rested the accuracy of information about the known world. Although confined to his monastery, he sought counsel from travelers, from written materials, from rumors, and from every source he could unearth. He wrote, as James Cowan notes in *A Mapmaker's Dream:*

> All we picture to ourselves or see with our eyes is inseparable from ourselves . . . I am ready . . . This final rendition of my map . . . must affirm the existence of the world I have discovered with as much honesty as I can muster . . . Decorations feature mermaids and dragons, together with a variety of strange animals . . . But it is hard to begin . . . Everything I now know is based on the perspective of others.

This sense of insecurity and apprehension when you do not know the subject at first hand is as true of the cloistered biologist as of the cloistered monk. There is no substitute for fieldwork—hard, humbling research in the field. Without it, like the blind men with the elephant, we see only a part of life's puzzle. If we do not understand the diversity of nature at a scale far beyond the local, then we limit our understanding of nature to the "perspective of others."

Doing ecological experiments is fun, for they provide results fairly quickly and meet with great acceptance in the scientific community. I went to Argentina initially to do ecological research as a part of my doctoral studies. But as I began my work in the Monte it became clear that few mammalogists in the country could provide a correct identification for many of the species I encountered. I once began a live-capture study of grass mice *(Akodon)* at El Infiernillo, the high grassland in Tucumán Province. Grass mice, you recall, constitute the complex group of species that look so much alike. Unfortunately, I was unable to number the individuals I was catching because I could not tell if I had one, two, three, or even more species. I was adrift like Fra Mauro and mermaids were beginning to appear in the edges of my map.

I had to terminate the ecological study because I could not identify the species. No one in Argentina could identify them either. Eventually I learned that no one in the world could give me the identifications of some of the animals I was catching because no one had studied them. No one had tackled this enormous assemblage of species that resembled each other greatly and that occurred throughout South America. Like the world of Fra Mauro, they were terra incognita. Unfortunately, I had a bunch of them in my traps. I decided that while ecological research was important and fun—and provided great intellectual rewards because of its experimental methodology—I was in a unique position to study the mammals of Argentina, to try to clarify the taxonomy of at least some of the groups I was encountering. If not me, who?

When I began to conduct field research that was designed to find new species, I was astounded at how poorly known the country's mammals really were. Now I consider myself lucky to have worked with a poorly known fauna. Many mammalogists trained in well-studied countries, conducting experimental research on well-studied species, are incapable of understanding the importance of foundational research on taxonomy, natural history, and distribution. Survey work and the task of chronicling nature's diversity receive little respect in the scientific community. They are activities thought to belong to another time. Being adrift in a country with a poorly known fauna was a life-changing experience for me, and so I turned my attention to the taxonomy, systematics, and distribution of the mammals of Argentina.

Had the taxonomic work already been done on the species I wished to study, I would happily have continued my ecological research. But the chaotic taxonomy of the mammals I came to love was unacceptable to me. No one had studied any of these species for any extended period; yet I was coming to know them well. I decided that if I just walked away from the challenge of these unknown species, it might be decades before anyone else would have an opportunity to study them. I felt a sense of duty—a powerful incentive to persevere.

Such thoughts were on my mind in 1984 when I presented a paper at a conference in Calcutta, India, highlighting the importance of survey research, field guides, and other publications that make basic data on species available to the public and to specialists. I argued that biologists have a duty to discover and describe nature's diversity, to study natural history, and to investigate the origins of this diversity. Thinking about this paper after it was published, I decided that my argument had merit. If I truly believed what I wrote, then I owed it to my discipline to develop foundational data on the mammals of South America. Someone had to make these kinds of decisions if the fauna of this poorly known region was ever to be clarified. This resolve led me down an uncommon path in biology, but one that has given me deep respect for the mammals of one of the most interesting areas on earth.

Much of the work I have published in recent years, along with my students and colleagues, will make the species we study more accessible to researchers in the future. Together we published guides to the mammals of Salta and Tucumán provinces, two areas in northwest Argentina that are especially rich in species. In addition, we published a guide to the bats of Argentina. Rubén Barquez became the world's authority on the bats of Argentina, while doing his doctoral thesis on the topic, and Ricardo Ojeda became the leading desert biologist in Argentina, being the first person to collect many of the previously undescribed animals that we named. Janet Braun has a better understanding of the taxonomy of the rodents of the country than almost anyone. Together we have published well over two hundred papers and several books on the mammals of Argentina.

I believe that this work will bear fruit in the future. Some time in the twenty-first century young mammalogists will arrive in the Monte, as I did thirty years ago. Unlike me, however, they will know a great deal about that desert's mammals, for those animals will no longer be unknown and undescribed. They will know that mammals are not abundant there. When they finally catch a mammal, they will know what species they have caught. They will be able to conduct an elegant ecological study on the mammals of that desert or of many other habitats in Argentina because the initial work has already been done. The species will have names. Something will be known about their natural history, distribution, habitat selection, reproduction, ecology, and evolution. They will no longer be anonymous and unidentifiable. We will have removed many of the mermaids and monsters from the margins of the map.

From Howling Wolf Mice to Fairy Armadillos

The desert. We now have many photographs that we use to fog that simple word. The days tumble together, the sun at noon annoys with light and flattens everything the eye sees into boredom. The ground boils with the goings of large ants, and every plant seems to rake the flesh with a lust for blood. If a careful tabulation is kept of good moments, they turn out to be very few. A half hour riding the cusp of dawn, a few minutes as the sun melts below the horizon. The seconds in the middle of the night when the body turns, the eyes briefly open, and the cold sky burns with white stars. That's about it.

And yet we go. We say, "We're going into the desert." We seldom say quite where because it does not matter. And besides, everyone knows where we are going. Into the desert.

Charles Bowden, *Desierto*, 1991

Life in the world's deserts is diverse, unique, and, in a sense, predictable from one desert to the next. But the ability to foretell the types of species that will evolve in each desert exists only at a basic level. Indeed, the word "foretell" implies prophesy, as if a desert existed that had not already been explored but that would contain types of species one could actually predict. A much better term is "postdiction," because one is attempting to discern the underlying forces that led desert mammals to develop the way they did. If we could discern and explain a pattern that was robust from one desert to another, this would indicate that we are able to understand why species develop the way they do, and how they come together to form communities of organisms in the world's arid lands.

As I expanded my research to the deserts of the Old World it became clear that the story of how mammals adapt to deserts was much more complicated than had been surmised. I was finding bipedal rodents, but they were not specialists on seeds. Instead, they shared a diet with the burrowing gophers of North America or the tunnel-dwelling, thumping tucu-tucos of the Argentine Monte. Things were not as they appeared to be before I began my intercontinental comparisons of desert faunas. First in Iran, then in Egypt, the same story was repeated. Being bipedal was clearly an important way for some small desert mammals to move through the desert, but only for a few species in each desert, and these species were not necessarily seed specialists. Moreover, as in North America, most of the desert species that *did* eat seeds were not bipedal. Rather, like the granivorous North American pocket mice, they ran around on all four feet.

Those species that specialized on seeds all lived in burrows, but so did most other desert rodents. Most seed eaters were active only at night, but that was true of almost all small desert mammals. Most deserts are simply too hot for a small mammal to stay aboveground during the day. Most seed specialists maintained extensive burrows where seeds could be stored for years at a time. Most did not need to drink water, obtaining their moisture through a complex array of physiological adaptations. Water conservation influenced everything from the anatomy of the nasal chamber (with inner webs of blood vessels that permit cooling exhaled air to recapture some of its moisture) to kidney structure (an elongate renal papilla to concentrate urine). The circulatory system might be modified to shunt blood to or away from an area where heat retention or dissipation was required. Small mammals also had a vast array of many other types of complex behavioral patterns that minimized water loss.

Certainly at the level of physiology—the physicochemical functioning of the cells and organs of the organism—strong convergence was evident. The physiological systems of the mammals had solved the problem of obtaining water and minimizing water loss in similar ways. Knut Schmidt-Nielsen had shown this in the mid-twentieth century and his studies are still germane. Physiologically, mammals converged upon each other in the types of adaptations that developed for survival in the desert. At the level of the whole organism, however, and especially at the levels of organization above the individual (communities of species, for example), convergent evolution became more variable in its manifestation.

Within each desert some species of rodents were bipedal and others were specialized for life under the broiling soil. Still others moved into rock piles, where green vegetation, occasional water, and cool, deep shelters were available throughout the year. If annual plants showed a propensity to wax abundant in deserts, so did cryptogamic plants, plants that spend the bulk of their life as bulbs or tubers buried under the desert soil. Unlike seeds, which escape heat and aridity by preserving the tissues from which a new plant grows within a watertight shell, cryptogams escape the heat by burying themselves in the soil until conditions aboveground are conducive to plant growth. In southern Africa, Bushmen forage on these plants for food and water.

Here was another food type waiting to be exploited by small mammals in deserts, and many species were quick to move into that niche. Some were specialized burrowers—gophers, mole rats, and tucu-tucos—but others were not. Surprisingly, one group that moved into this food niche in Old World deserts hopped on its hind legs like the kangaroo rats. The various jerboas I encountered were, in a sense, bipedal gophers and bipedal tucu-tucos. They ate roots and tubers, but instead of burrowing underground in search of food, they hopped about on the surface as they foraged.

In the late 1980s and early 1990s I extended my analyses to desert rodents living in all of the world's deserts. I had shown earlier that the morphology of desert animals strongly reflected both their diet and their ecology. I analyzed museum specimens from collections throughout the world and examined the increasing number of natural history papers that had been published on species that had hardly even been considered in the past. The pace of desert research outside the United States had quickened, and the published papers contained precisely the types of information I required in order to discern patterns of desert rodent adaptation.

Since most research on desert mammal ecology and physiology had been done on North American species, there was a clear bias in the scientific literature that all deserts should be like those of the United States. Nevertheless, although all desert rodents did not follow the model of desert adaptation exemplified by North American kangaroo rats, many characteristics appeared among desert species with great repeatability.

First is the habit of hopping on the hind legs. Bipedal rodents are surely ecological analogues from one desert to another in the way they move. Moreover, if we exclude kangaroos and their relatives, bipedal locomotion is common only in desert rodents—out of more than two thousand species of rodents in the world (almost half of all mammal species), only a few dozen are bipedal, and almost all of these are restricted to arid or semiarid areas. Far from being an adaptation for gathering seeds, however, bipedalism appears to be an adaptation for escaping predators. Species that are bipedal are much speedier and harder to catch than others. I found the Old World desert species living in areas where almost no plants grew, yet the animals were nearly impossible to catch as they bounded rapidly across the bare ground.

Similar diets can result in remarkably similar adaptations. If a desert rodent develops the ability to eat the leaves of saltbushes, for example, it will also develop unusual incisors to strip the leaves of their coating of salt crystals. Mammals cannot eat the salt crystals without perishing, so they must find a way to remove them from the plant. Three different species in three different deserts have done this. Each has converged in details of the teeth and kidneys. They may differ in exactly how the leaves are stripped, but they have solved the problem of using saltbush leaves for food and water in remarkably similar ways, and their level of convergence is pronounced.

Whether or not a desert rodent harvests a challenging resource such as saltbush, however, the similarities in various aspects of the biology of desert species can be striking. Most desert rodents have kidneys whose structure is designed to conserve water, and the gross morphology and ultrastructure of the kidneys are similar. Most have developed the ability to minimize water loss by producing dry feces. They can recapture water that would otherwise be lost in the air they exhale by using long nasal bones along which the warm exhaled air condenses (so they have a long snout).

They live in sealed burrows during the heat of the day, and most can live without ever drinking free water. The chambers surrounding their inner ears are enormously inflated to capture the slightest sounds on the desert air. Desert rodents have large eyes located on the top of the head like turrets, well positioned to detect predators' attacks, which can come from any direction in the sparsely vegetated desert. Their fur is pale, usually blond above and white below—a color scheme that makes them almost impossible to see in the dim light of evening as they move across the pale desert soils. They have a long tail and, if they are bipedal it will often have a tuft of hairs that is mostly black but has a starkly contrasting white tip. The black and white tuft may help deflect the attack of a predator toward the bright, moving tip and away from the animal's vulnerable head.

There are other ways to exist in a desert besides being bipedal. If a species has opted for an underground existence it will have small eyes, or may have entirely lost the ability to see. Such species will have a fusiform body, a shape that permits them to move easily within narrow tunnels. The external ears, the pinnae, will be small or missing altogether, the forefeet will be large, and the incisors will be thick and broad, since they are used for digging through the hard desert soil. Burrowing species will usually eat roots, tubers, and other plant parts, but not seeds. Some may eat invertebrates and small animals.

Once a major life mode has developed—burrowing or bipedal hopping, for example—many other characteristics follow. All in all, desert rodents are among the most predictable of all small mammals from the point of view of adaptations to survive under arid conditions. The desert is a harsh place for a small mammal to exist and there are only so many ways to exploit it.

Most species are nocturnal. This may have necessitated the development of an acute sense of smell and lightning-quick reflexes. Each night the rodents overcome the challenges posed by owls, snakes, foxes, badgers, skunks, and other predators. Each animal that prowls the desert tonight is the descendant of all of the members of its lineage that have come before. Beyond that, each individual is the descendant of only the most successful desert animals, those that lived long enough, that were wily enough, quick enough, or so highly adapted to their desert habitat that they were able to leave more offspring to carry their genes forward into succeeding generations. The rodents that are foraging for seeds in the world's deserts tonight are the apexes of adaptation of their entire lineage back to when their ancestors first encountered aridity.

As I expanded my research to include specimens from all deserts, I was able to consider how the adaptations of mammals are similar among species that are only distantly related. For example, the small mammals of the Monte and Sonoran deserts had proven less similar than those of the Sonoran and Iranian deserts, even though the faunas of the two New World deserts were more closely related evolutionarily than were the faunas of the Iranian and Sonoran deserts. When the mammal fauna of the Sahara Desert was considered, its mammals were much more similar to those of the Sonoran Desert, ecologically and morphologically, than to those of the Monte Desert of southern South America, though not as closely related genetically. This means that the desert faunas of the Northern Hemisphere had converged upon each other, but the mammals of the Monte Desert in Argentina had developed in a different direction—at least as far as "classic" bipedal desert rodents were concerned. This was strong evidence for the force of convergent evolution, although the exceptions were interesting.

My analyses showed that there is an element of chance in the evolutionary selection of the paths toward desert adaptation taken by rodents. I found that being bipedal was not related to diet. There were bipedal seed eaters, leaf eaters, root and tuber eaters, insect eaters, and even omnivores.

I examined patterns in the evolution of desert rodents from throughout the world and noted that adaptations to deserts were the result of an evolutionary deck of cards being shuffled and dealt to different species in different deserts. However, nature was playing with pretty much the same deck of cards in each desert. A small mammal striving to survive in the desert would have to play a hand with these desert-adaptation cards, and its options for becoming a desert specialist were limited. Its kidneys had to concentrate urine and minimize water loss. It had to live in a burrow. If it was to be successful, it had to solve the challenges of heat, aridity, sparse vegetation, and the many other evolutionary challenges facing a mammal trying to live in a harsh, arid climate.

When convergent evolution is viewed in this manner, it might be expected to result from similar cards being dealt to many organisms in each desert. But the evolutionary deck is not limited to fifty-two cards. Rather, the cards are the tens of thousands of genes that provide the blueprint for each species. Surely many of these genes are associated with each other, either through proximity along the chromosome or through related functional effects in the organism. For example, all bipedal species have small forelimbs. Could this be due simply to chance? Could the long, white-tufted tail that is common among bipedal animals also be due merely to chance? Possibly families of related genes are inherited as groups that influence entire suites of traits at the same time. Subsequently, selective pressures on individual genes or groups of genes for each trait may be strong enough to result in similar form and function developing among distantly related animals.

Given the limitations inherent in the genes of mammals, there are only so many options that can present themselves for species attempting to live in a challenging environment. Even though this enormous deck of cards is similar from desert to desert, it is not exactly the same, and the shuffling that occurs through mutation, recombination, and natural selection ensures that the hands of cards that are dealt are different for different species, as well as for different deserts.

There are constraints established early on in the development of a desert's fauna that will be carried forward through time. For example, the jerboas of the Old World deserts, which may eat anything from roots and tubers to insects or seeds, are able to hibernate during the long, cold winter. Did the ability to hibernate free them from the necessity of storing food? Perhaps, for no jerboa stores grain or other vegetation like a kangaroo rat. But kangaroo rats do not hibernate. In this case, the genetic constraints that may restrict hibernation may also limit the evolutionary options of the rodents that make up the desert's fauna.

Thus in both New and Old World deserts the evolutionary cards for bipedal locomotion were received by a small subset of the species evolving in each desert, although the cards dealing with the diet of the animals and with their physiological ability to hibernate were different. In the Monte Desert, similar cards were also dealt to small mammals, especially a few hundred thousand years ago when animals very similar to jerboas lived in the desert, but the mammals that received these cards were marsupials, not rodents. Their fossilized remains tell us that they were desert species as highly specialized as a jerboa or a kangaroo rat. In the ultimate evolutionary game of chance, these bipedal desert marsupials went extinct for reasons that may never be understood.

As in all evolutionary stories, chance plays an important role. It seems clear that even the great dinosaurs were subjected to the sudden snuffing out of one of the greatest vertebrate lineages that ever existed by the chance occurrence of an asteroid colliding with the earth. Nothing in their evolutionary history could have prepared them for such an unlikely event. Without the ruling reptiles, the stage was set for a different group of organisms, mammals, to move into ascendancy. Had it not been for the asteroid, I would probably not be studying desert rodents or writing these words, and you would not be reading them. The dinosaur deck of cards was removed from play, and a new game, mostly with new cards and new players, began.

At the level of interacting groups of populations in communities and ecosystems, convergent evolution is more difficult to detect. The card analogy makes clear that similar traits are found among suites of species that have evolved in different deserts. But it is almost impossible to predict how many species, or what types of species, will characterize each fauna.

In the deserts of North America are several species of grasshopper mice (*Onychomys*), 2-ounce predators that feed on invertebrates and vertebrates and fill the desert night with their tiny howls, which account for their also being called wolf mice. No other desert rodent preys on other rodents. There are predators in all deserts, of course, but thus far no ecological equivalent of the wolf mouse has been found. Thus any community with a wolf mouse will have an element of uniqueness—a singular niche type that is found nowhere else. Such incomparable species, in the truest sense of the term, appear in each desert, whether they are the ferocious wolf mice of the United States or the pink fairy armadillo that swims through the sands of the Monte. These unique species give each desert a special flavor, as surely as secret spices make a master chef's dish distinctive.

Aridity's Cornucopia 16

There is a popular belief that the camel is about the only mammal in the desert. But the reality is very different, for a surprisingly broad spectrum of mammal groups has successfully tackled the problems of desert survival. Most are comparatively secretive: only a large animal, almost immune from attack, could afford to be seen against such a starkly revealing habitat . . . A walk or drive through the desert at night reveals a startling number of pairs of watching eyes of all sizes, set in shadowy, swiftly retreating bodies.

Jim Flegg, *Deserts: A Miracle of Life*, 1993

What do deserts mean to the diversity of life on earth? When the popular media report daily on the plight of the rainforest, why should the fate of the world's deserts be of any importance?

The 1980s and 1990s were the decades of the rainforest, for during this period the great tropical forests were recognized as being both especially rich in species and greatly threatened by deforestation. As both the diversity and endangerment of tropical forests were described, a popular movement for their preservation began that was global in scope. The largest rainforest, the Amazon, was described as the "lungs" of the earth, the oxygen-producing capacity of its trees and their simultaneous consumption of carbon dioxide being analogized to the gas exchange that takes place in our own bodies. In many ways it was an apt analogy, but the story of diversity is much more complex than sheer numbers of species might suggest.

There are at least two components to diversity. One, and perhaps the simplest, is the absolute number of species that are supported by a habitat. The wet tropics are rich in species, but does this number supply all of the vital information that is required in order to make the hard decisions that must be made to conserve the species in particular regions? Is diversity merely the number of species in a given area? Many have argued that a major reason for not allowing species to go extinct is that their genetic uniqueness would be lost forever. There is much to be said for genetic uniqueness.

Rainforests support many species including, among mammals, many primates and bats. The lowland tropical forest is extremely rich in plants and insects, especially beetles. Many rainforest species are closely related, however. That is, in the case of many, if not most, organisms, tropical species cluster together in species-rich genera. Each species is adapted for life in the tropical forest and each is genetically unique. But genetically speaking, those species that are clustered within a speciose genus are not as distinctive one from the other as are sets of species that belong to different genera. Thus a zebra and a rabbit contain more unique genetic information than do two species of zebras or two species of rabbits.

Think of the Noah conundrum. Noah's mythical Ark, for all of its capacity in cubits, could not hold all of the world's animals. Like a modern wildlife manager, Noah had to choose. How does one maximize diversity in choosing to save sets of organisms? If you had to choose and could save only twenty-six species of mammals, which would you select? It is doubtful that you would save twenty-six species of Amazonian monkeys. It is similarly doubtful that you would save twenty-six species of North American field mice *(Peromyscus)* or South American grass mice *(Akodon)*. Instinctively, you would recognize that twenty-six species of *Peromyscus* contain a variety of unique genetic information specific to field mice belonging to the genus *Peromyscus*, but there is little of the whale or tiger in a field mouse. If you saved twenty-six species in any single genus, most of the rich variety of mammals in the world would be lost.

Very likely, when faced with this dilemma, you would elect to save a species in each of the twenty-six orders of mammals. This would preserve an elephant, a whale, a primate, a rodent, an aardvark, a hyrax, a pronghorn, a lion, and so forth. In essence, you would be maximizing the genetic diversity of the surviving species, not merely saving a particular number of species. Genes are packets of information and contain all of the information required to build, replicate, and maintain members of a species, but all species are not equal in the evolutionary game. The information contained in two closely related species is more similar than the information contained in two species that are in separate genera. Similarly, two species in different families contain an even greater amount of unique genetic information. It takes more information to make a tiger and a mouse than it takes to make two mice in the same genus.

What does all this have to do with deserts and biodiversity? Scientists like to simplify when trying to communicate an idea to the public. The rainforest has many species. It is beautiful, rich, green, wet, and even sacred in a sense. If scientists can gain the support of the public in averting an ecological tragedy like the loss of the rainforest, they will have taken a big step toward slowing species loss. The romance of the rainforest, coupled with threats to its continued existence, has galvanized people all over the world. Jacques Cousteau made many trips to the Amazon, Vice President Gore visited the forest, and the community of ecologists has kept up a drumbeat about species loss in the rainforest. Even Hollywood has gotten on board, with movies such as *Medicine Man, At Play in the Fields of the Lord*, and *The Emerald Forest*, that highlight the diversity of life in the rainforest.

What is said about the rainforest is true, but it is not the only truth. The rainforest is rich in species, but it is not the only cradle of diversity, or even the most important one. From the point of view of South American mammals, there is greater diversity outside the rainforest than there is inside. I showed this in an article in *Science* magazine in 1992 that elicited some exceedingly critical responses. What I showed was that when all of the arid and semiarid habitats of South America are combined and compared to the Amazon rainforest, there are more species, genera, and families of mammals in the arid and dry areas (loosely defined) than there are in the rainforest. Similar work on birds and spiders has supported my hypothesis.

How can habitats that are dry be more diverse than the tropical wet forest? For one thing, the arid and semiarid areas of South America are somewhat more extensive in area than the Amazon rainforest. There is a rough rule in ecology that larger areas support more species and more different types of organisms than do smaller areas. When I controlled for total area, the number of species in the drylands was about the same as (though slightly larger than) the number of species in the rainforest. More important, the numbers of genera and families were higher in the drylands than in the rainforest. This means that there is more unique genetic information in South America's arid lands (including grasslands that are subject to extensive droughts) than in the rainforest.

In addition, the Amazon harbors very few species that are limited to (endemic to) the rainforest—many also occur in savannas, grasslands, semiarid thorn scrub, and other habitats. A good example is the jaguar *(Panthera onca)*, the great cat of South America. Jaguars live in the Amazon, but also in the Brazilian Cerrado (a dry savanna), the semiarid Caatinga (a tropical thorn scrub), the Chacoan thorn forest, and other habitats from northernmost South America to northern Argentina. In historic times, jaguars even extended into the Monte Desert. In fact, they still range as far north as the southwestern United States, where they have recently been seen in southernmost New Mexico.

Most of the mammal species that are restricted to the Amazon are monkeys and bats, and a few rodents. There are relatively few endemics, however, which means that there are few species whose total distribution is limited to that habitat. This has significance for conservation. If the Amazon forest is turned into a parking lot, there will still be jaguars in North and South America, for they are not restricted to the lowland rainforest. One can do similar calculations with each species, which I did, and the surprising result is that the Amazon is not the greatest repository of the genetic diversity of South America's mammals. The case for the distinctiveness of the forest as a zone of mammalian uniqueness has been overstated.

Suppose that Noah really did exist and there really was a great flood. Suppose God decided that only one major habitat type could be saved (wet or dry, to make it simple), whereas all the rest would be lost to the floodwaters. If God told Noah to choose the habitat from which he could save all of the mammals from extinction, what choice should Noah have made? Most people will say, "Save the mammals of the rainforest, for it is the richest habitat." My research made clear that most people would be wrong. In order to be remembered as the savior of mammalian biodiversity in South America, Noah would have had to say: "I have decided to save the mammals of the South American drylands. I will select those species that live in the Chaco, the Monte, the Caatinga, the Cerrado (a vast tropical savanna), the pampas (the grasslands of Argentina and Uruguay), the puna (the high Andean desert), and the páramo (the high Andean tropical grasslands)."

By deciding to save the species in the drylands Noah would have made the correct choice. That difficult decision would have saved much of the diversity of the Amazon as well, because so many of its species are shared with other, drier habitats. But the decision to save the mammals of the drier habitats would also have preserved the unique families, genera, and species that have adapted to aridity over great spans of time. Many rainforest mammals are found in dry scrublands or thorn forests, or even in arid areas, but desert animals are not found in rainforests. Thus deserts support a fauna that is more distinctive, one that is not replicated in wetter habitats. In the deserts of South America, for example, entire families of mammals are limited to arid and semiarid habitats. Such families as the Camelidae (the guanaco and vicuña), Chinchillidae (chinchillas, vizcachas), Ctenomyidae (burrowing tucu-tucos), Caviidae (guinea pigs, maras, cavies), Octodontidae (vizcacha rats), Abrocomidae (chinchilla rats), and others occur only in arid areas.

If only the mammals of the rainforest were saved to the exclusion of the mammals found in the drylands of South America, the mammal fauna of the South American continent would be poorer indeed. Gone would be the Chacoan peccary, most armadillos, and many other species found only in areas where drought is extensive. The pressures of natural selection on species faced with the challenge of aridity are so great, and the consequent adaptations so exceptional, that the drylands are enriched by sets of species, genera, and families that occur nowhere else.

Perhaps those who were critical of my research were simply distracted by the fact that the simple measure, species number, was also higher outside the rainforest than within. This finding was unexpected and was viewed with alarm and dismay by those whose constant repetition of the mantra of the unparalleled diversity of the rainforest bordered on hyperbole. The differences in total numbers of species between rainforest and arid land mammals were small, and new data or a refinement of habitat limits could show that the rainforest supports more species. However, the data on genera and families are more robust, and these numbers will not easily be invalidated. The rainforest is less diverse at higher taxonomic levels, indicating that the species of the drylands are more important from the standpoint of their unique genetic natures. Certainly they have had to overcome greater hurdles in adapting to life in their environment.

The desert world is rich in unique and unreplicated forms of life. In some ways deserts seem as different from tropical habitats as other planets are from earth. The species of the harsh and challenging world of aridity have had to develop genetic codes that permit their continued existence in habitats that would otherwise defy habitation. The world's diversity of genetic uniqueness would be much poorer had desert organisms not evolved. Similarly, if life in the desert were lost owing to lack of concern for the persistence of arid habitats and their species, entire families and genera would disappear into the dark void of extinction. We are seduced by the green world of the tropics and think little of the rich life of arid regions. No movies describe the interesting life of deserts. No television specials cover the rich life of the arid world. The prevailing view was spoken long ago by Alec Guiness as King Faisal in *Lawrence of Arabia*, "There is nothing *in* the desert. No man needs nothing." We view life in the arid places as harsh, dangerous, and hostile. It is the life of the xeric world, however, that rings with the unequaled tonalities of starkly different taxa—peerless species, genera, and families of mammals that provide a rich alternative to the thinly sectioned mammalian diversity of the tropics.

Life in the Desert of Salt

We sat on a crate of oranges and thought what good men most biologists are, the tenors of the scientific world—temperamental, moody, lecherous, loud-laughing, and healthy . . . The true biologist deals with life, with teeming boisterous life, and learns something from it, learns that the first rule of life is living . . . Your true biologist will sing you a song as loud and off-key as a blacksmith, for he knows that morals are too often diagnostic of prostatitis . . . he is very good company, and . . . does not confuse a low hormone productivity with moral ethics.

John Steinbeck, *The Log from the Sea of Cortez*, 1951

I was able to visit northwestern Argentina again in 1999. Although my colleagues and I had worked in the region off and on for over thirty years, I was sure that there were still many new organisms waiting to be discovered. In planning the trip I perused many maps, picturing each area in my mind's eye whether or not I had ever been there. As I stared at the map, my mind returned once again to the isolated Bolsón de Pipanaco, the "lost" valley near Andalgalá in Catamarca Province where my work began so long ago. In that valley is a large salt flat—*salar*—that extends more than 50 miles from north to south and that is up to 10 miles wide. The salar had always been inaccessible to me because exploring it required a four-wheel-drive vehicle or a long trip on horseback, neither of which had been possible in the past. But our research grant in 1999 was large enough to cover the purchase of a new four-wheel-drive pickup that would finally allow us to explore the salar.

As I looked at the map the history of the area flashed through my mind. I saw the ancient movements of habitats and faunas; the mountain chains thrusting upward; the inflow and outflow of the Chacoan thorn scrub as it rushed tidelike into the valley during moist times and retreated during arid periods, leaving the Monte behind; and the formation of the salt lake itself. Thinking back over our explorations of the mammals of the Monte, I realized that there was a high probability that a mammal that was specialized on the halophytic vegetation of salt flats could have become isolated in the valley by the rising mountains and retreating forests. Since the mountains arose mainly over the last five to ten million years, the rodent (I was sure it would be a rodent) could not be a sigmodontine. It would have to be a caviomorph. And one of the ancient families of caviomorphs that was associated with aridity and with lowland desert was the Octodontidae, the vizcacha rats.

The fossil record told me that caviomorphs, and especially octodontids, had been in the desert since it first formed. The Pipanaco Salt Flat (Salar de Pipanaco) was the right age to have been present before the mountains arose in the late Pliocene. To the south lived the plains vizcacha rat, *Tympanoctomys barrerae*, which had developed the little toothlike brushes near its mouth that it used to strip the saltbush leaves of their salt covering. I did not know exactly which salt-loving plants would grow near the Salar de Pipanaco, but I did know that when you look down on the salt flat from the surrounding mountains you can see some sort of low vegetation around the salar. It had to be a ribbon of salt-loving plants. The stage was set for the evolutionary play. All of the actors seemed to be in place. What did my instincts tell me about what might be found in this area that had never been visited by a mammalogist?

I spoke with Janet Braun and wrote to Rubén Barquez and Mónica Díaz, colleagues sharing the grant. I told them that I was planning a trip that would take us into the Salar de Pipanaco. I thought we would find a new genus of salt-specialized rodent living in the halophytic vegetation near the salt lake. It would belong to the family Octodontidae. Its nearest relative would be *Tympanoctomys*, the plains vizcacha rat. We would first look for the unusual mounds of the animal and after finding one would probably have to excavate it to catch the rodent, since the rats would be hard to trap. Perhaps it is a tribute to our friendship that they believed me.

There is no easy way to enter the Salar de Pipanaco. There are a few dust-blown trails into the mesquite forest surrounding the salt lake that are used by woodcutters, but they are very treacherous roads. The inner basin of the valley is covered with a deep, powdery dust that makes it difficult to drive without getting stuck. Moreover, if you do not know the way in, it is easy to get lost amidst the thorn trees. Our first attempt was aborted after about 10 miles of wandering through the trees. The thorny branches mercilessly scraped the beautiful truck, our first new vehicle in three decades of research. We managed to return to the main road, then found another trail leading into the valley.

Gradually we worked our way through the tall mesquite forest. We stopped to talk to a man who lived in a small house near the main road. He said that he knew the area well and was familiar with most of the fauna. When we described the type of animal we were looking for, he said no animals like that lived around the salt lake. We thanked him and headed on into the valley. Forest gave way to low shrubs. Salt deposits began to appear on the soil surface. We were nearing the salt flat.

The Pipanaco salt flat in Catamarca Province, Argentina, seen through our windshield as we set out to search for a new genus and species of rodent. (Photo by M. A. Mares.)

The vegetation became sparser and more stunted. Finally the shrubby halophytic plants appeared. These were not saltbush (*Atriplex*), but a different genus of halophyte (*Heterostachys*) that we had encountered in other salt lakes to the south. It was a plant that, unlike saltbush, did not have gray-green leaves. In fact, it appeared not to have leaves at all. Its green branches seemed to be composed of series of diminutive blue-green balls. Each ball was made up of masses of tiny leaves tightly pressed together. The leaves were very salty to the taste. *Heterostachys* is a member of the family Chenopodiaceae, the goosefoot family, the same family as saltbush and so many other halophytes that grow in salt deserts throughout the world. It is a plant that grows in extremely salty areas.

As we drove farther into the area encircling the salt flat the soil surface became covered with a crust of snow-white salt. Blue-green patches of the salt bushes punctuated the stark white landscape. We stopped and took a long walk through the low vegetation. We found armadillo burrows and tucu-tuco diggings, but no large mounds that would characterize the animal I was seeking.

We crossed the salt flat proper, following powdery gray tracks sunk deep into the blindingly white salt. The true salt flat had no vegetation at all, only a hard crust of salt that was impervious to plants, even the chenopods. We crossed the northern edge of the salar at its narrowest point and again entered the narrow strip of halophytic vegetation. Beyond the low plants we could see the desert scrub proper growing just outside the limits of the salt deposits. We stopped again for another walk through the halophytic desert.

Within minutes we found the mounds we were seeking. They were unlike those of the plains vizcacha rat, but were clearly mounds of a rodent. I could see that it ate the salt plant *Heterostachys ritteriana*, which was the only plant that grew on the mounds. The signs of nipped branches were everywhere. The mounds we had discovered were not as large as those of the plains vizcacha rat, each of which was about 30 feet by 10 feet. These mounds were only 12 feet by 6 feet, but they were complex structures. Moreover, there were up to six mounds placed together in isolated clusters, with a larger mound being connected to several smaller ones. Could a single animal live in so many mounds? Most of the mounds showed no signs of recent activity, but a few—a very few—did show evidence that an animal had been there not long ago. We grabbed our shovels and began to dig. The digging was difficult, and after opening an entire cluster of mounds and moving more than a thousand pounds of soil we had found nothing. We began to excavate a second set of mounds.

The mounds of a previously undescribed genus and species of rodent in the Pipanaco salt flat in Catamarca Province, Argentina. The ground surface is coated with salt. (Photo by M. A. Mares.)

It is a strange feeling to be searching for an animal that you know exists, but have never seen. I doubt that anyone had ever seen this reclusive rodent living in a small patch of salt in an isolated desert valley. We had really moved into the realm of cryptozoology—the search for unknown forms of life. In a way we were seeking our own variant of the Yeti—a tiny Sasquatch of the Bolsón de Pipanaco—hoping to find a new animal that until now had lived only in my imagination. What would it look like? I had a picture in my mind, as did we all, but we did not know for sure. No one knew. We were trying, literally, to unearth a species and a genus that I had imagined months before on another continent 6,000 miles away, a genus and species that had inhabited this isolated valley in absolute anonymity for millions of years.

The second cluster of mounds, like the first, had many openings. As with the plains vizcacha rat years before, we plugged the burrow openings with live traps, expecting at least one animal to race away from the mound in panic and run into one of the waiting traps. We continued to dig, but no traps slammed shut. There were burrows at several levels and the burrow system was complex, with many intersecting passages and large chambers where tunnels intersected. We uncovered food stores, latrines, and central chambers located amidst the roots of the salt plants that grew on the mound. We excavated most of this set of mounds too, but found no rodent.

We proceeded by picking a tunnel and following it from the central part of the mound to the trap that we had placed just outside the tunnel's exit from the mound. As each tunnel was opened and cleared, we moved to the next. There were about thirty tunnels in the mound complex, some placed 3 feet underground and woven between the tough roots of the salt plants. Finally only a single tunnel remained. The animal had to be there. All of us held our breaths as we waited for it run into the last remaining trap. We cleared away the last bit of sand, and I lifted the last trap—no rodent.

"Nothing," I said.

Then suddenly Janet screamed, "There it is!"

We watched transfixed as a large golden rodent about the size of a woodrat and with a bright red bushy tail leaped from the sand and tried to escape. Mónica dove down and grabbed it. This rodent was beautiful. We all leaped with joy and hugged each other. We had found it! As we examined the animal, it became obvious that it belonged to no genus or species that had ever been described. It was, indeed, an octodontid. We had found the salt rat of the isolated Valley of Pipanaco—an animal that had been isolated so many millions of years ago that it was generically distinct from its few relatives in the family Octodontidae. It was yet another unique life form living in the Monte Desert.

Our work over the next several days involved mapping burrows, examining the animals in greater detail, collecting plants, and surveying the area. The animal was clearly not in the same genus as the plains vizcacha rat *(Tympanoctomys)*. It had two sets of the little toothlike brushes of that species, but those of the new species were less pronounced and less stiff. The new mammal ate the little halophytic plant *Heterostachys*, not saltbush. Its mouth was small and round, ideal for devouring the tiny leaf balls of *Heterostachys*. And although the new mammal looked somewhat like the brush-tailed vizcacha rat *(Octomys mimax)*, its teeth were very different from those of that species.

The golden vizcacha rat *(Pipanacoctomys aureus)*. (Photo by M. A. Mares.)

We determined that the new species was exceedingly rare in the immediate area, although we still do not know how far it extends along the edge of the salt flat. Its population density is low over most of the area, although in a few locations the rats are fairly abundant, with up to twelve per acre of prime habitat, a very dense population for a mid-sized small mammal in a salt desert. We also learned that this rat has 92 pairs of chromosomes, among the highest number recorded for a mammal, although not as high as that of its cousin, the plains vizcacha rat, which has 102 pairs of chromosomes.

The discovery of an animal radically different from all other forms of life brought a flood of feelings, including anticipation, joy, elation, and even humility. We all knew that we had shared a moment that would be indelibly imprinted on our psyches. Few biologists discover new species of mammals and even fewer happen upon new genera. The fact that we had come to this isolated spot precisely to search for a new life form made our find especially satisfying.

Over the next several months we studied the specimens, confirming that this was, indeed, a new genus and species. We returned to the study area three times to examine the ecology of the animals. We learned that, for the most part, individual animals live alone in small burrow complexes. In at least one location, however, we encountered a massive agglomeration of mounds that extended over an area 200 feet long by 30 feet wide. It appeared to harbor dozens of these new vizcacha rats living in some sort of colony. We prepared the scientific paper that would provide the official description of the animal and give it a name. We decided to name the genus after the valley in which the rat lived, and the species after the unusual golden color. It would come to be known as *Pipanacoctomys* (the octodontid rodent from the Bolsón de Pipanaco) *aureus* (golden), the golden vizcacha rat.

As we were putting the finishing touches on the scientific paper, I began to think about the area where we had found the new mammal. Clearly, as all of our morphological analyses showed, it was related to the plains vizcacha rat, which lives far to the south. Yet, as even a cursory glance at a map indicates, there are a number of major salt flats between the known areas of occurrence of the two salt-loving vizcacha rats. I had crossed one of these vast salt flats several times in the past, but had never collected mammals there. I knew that no one had ever bothered to collect mammals in the area, for it did not look very inviting. It was a difficult place to work and was surrounded by dense Chacoan thorn scrub. Few people would go there willingly.

I was convinced that another new salt-loving mammal had to exist in the great salt flat known as the Salinas Grandes. I also knew that it would not be easy to get into the area, for there are almost no roads of any kind leading into that vast expanse of salt, occupying more than 2,000 square miles. The plains vizcacha rat had evolved in the southern salt flats and the golden vizcacha rat had remained isolated in the Pipanaco Valley far to the north. Surely there had to be another octodontid rodent specialized for life in the salt in Argentina's largest salt flat, located between these two points. I scanned the maps of the area and planned the next expedition.

Working in a salt flat is hard. Working in the Chaco is even more demanding. To get to the Salinas Grandes, we would first have to go through the daunting Chacoan thorn scrub. We would then have to find our way, somehow, across the trackless salt and look for an animal that, once again, existed only in my imagination. My colleagues and I began our trip in the Argentine winter (summer in the United States) of 2000. I chose to enter the salt in the winter because it is the only time that you can drive across the salt and not expect to lose the vehicle. In the summer during the rainy season, the salt lakes become flooded. The salt and sand combine to form a kind of quicksand, capable of swallowing a vehicle whole. Moreover, in the summer, when temperatures are high, the salt lakes, particularly those in the Chaco, become so hot that they can kill. You get burned from above and below as the snow-white salt reflects the searing sun. A Chacoan salt flat in summer makes the Chaco forest seem like a vacation spot.

We did not know how to get into the Salinas Grandes. Our maps showed only a few tracks tickling the edges and then disappearing into nothingness. I decided that we would approach it from the north, where a road seemed to reach to within a few miles of the salt. Perhaps we could find trails there that we could follow to reach the salt flat proper. We loaded our little truck with extra water and headed for Argentina's great salt flat.

It took two attempts to find the dirt road we were seeking. The winter cold was biting, but we vastly preferred that to the blistering heat of summer. Eventually, we reached a tiny settlement where the map showed the road ended. We asked a man there how to reach the salt. He thought us very strange to be seeking a way into the salt, but he pointed us down a dirt track that had been built for a single electric line that extended to a one-room schoolhouse about 8 miles away. "If you reach the school, you will find someone to tell you how to get to the salt." This was a somewhat unsettling comment, but we headed down the dirt track.

We arrived at the school late in the day. There was no one there. Nearby was a house that belonged to goatherds. There were goats and goat droppings everywhere. The stench was overpowering. The woman of the house, oblivious of the odor, greeted us and pointed to a track behind her place that led into the salt flat. She said that if we went south for several miles, we would find the shack of another goatherd. Past that point lay the open salt. We started down the track, but within a mile I decided that we should make camp. I did not want to enter an unknown salt flat in the dark.

In the morning, Mónica and Rubén stayed back to watch the camp while Janet and I went south to explore the road into the salt flat and to look for signs of an unfamiliar mammal. We reached the last goatherd's shack, but no one came to the door as we rolled slowly past. There was a faint trail extending beyond the shack, so we followed it. Within a few miles we crested a sand dune covered with cacti and thorn trees, and there before us lay the Great Salt Flat. We could see that once we crossed the dune, we would enter an area almost devoid of vegetation. In the distance we could see habitat islands rising above the white salt, looking for all the world like islands in a calm white sea. They seemed to be covered with the same kind of thorny vegetation that was on the dune where we had stopped. We decided to walk along the dunes that surrounded us before entering the salt. We found signs of many types of mammals, including armadillos and various rodents, but nothing that looked like the diggings of a salt-specialized octodontid.

I was uneasy about heading into the salt. It was an area that I did not know, although I was familiar with stories about people who had lost their vehicles in similar habitats in Argentina. We could not afford to lose the only new truck we had ever had, and I did not want us to end up mired in quicksand with no way to escape. Both events seemed likely as I looked out across the vast white nothingness that lay ahead. I decided to walk into the salt flat for several hundred yards to test the salt and see if it would support a vehicle. It seemed solid. We got into the truck, left the dune, and struck out across the salt toward the distant islands.

The farther we traveled into the salt flat, the stranger the habitat became. The islands, some 10 to 15 feet high, supported many of the plants that we had seen on the "mainland" near where we had camped, thick spiny scrub vegetation, as well as many candelabra-type cacti. But there were also plants

growing on them that I had not seen before. On the islands we found literally thousands of armadillo burrows as well as mounds made by tucu-tucos.
There were also a few burrows of animals we could not identify, but nothing
that looked as if it had been made by a rodent.

We checked island after island, stopping the truck and walking the perim-
eters and the elevated portions of each. Nothing. Eventually we reached a
point where the salt seemed to stretch to infinity. I knew we were looking
southeast toward the far reaches of the salt flat, across more than 100 miles of
salt and sand. I did not want to enter that trackless, dangerous area with only
two people and a single truck. At any point the solid ground below us could
suddenly give way to quicksand from which we could not escape, and I knew
that such a change in the terrain would be invisible from the truck. The surface could change very quickly. Looking south, I could see a large island appearing as a blue-green shadow through the glaring white air. Distances
across the salt are deceptive, for the wind can whip up salt crystals and sand
and make things appear farther away than they really are. Alternatively, when
the air is absolutely clear, distant objects can seem close. The island we were
seeing did not seem too far away. To reach it we had to launch ourselves into
the void, which we did.

We were following the faint track of another vehicle that had entered the
salt at some point during the last several months. But even with this trail to
follow it was nerve wracking to drive a truck across an unfamiliar salt flat. Everywhere water lies just a few feet below the surface. In some places it rises
so close to the surface that the salt and sand become liquid, although a thin,
surface crust remains, a trap for the unwary. I saw places where the color of
the salt changed from white to pale brown. In others it became more blindingly white. Could these areas indicate a change in the water table? We had
to put on sunglasses to minimize the glare, but I was worried that they would
not permit me to see subtle changes in the surface that might warn us about
an impending plunge into the mire, so I took them off again. I would drive
for a while, and then walk for a while. In this manner we finally reached the
large island we had seen in the distance. We began the now familiar task of
walking and looking.

Within a few minutes we came upon some strange burrows. Tucu-tucos or armadillos did not make them. We had seen a population of cavies (*Microcavia australis*) at the last goatherd area, but these burrows did not appear to have been made by cavies. Snakes, lizards, tortoises, birds, amphibians, tarantulas and other spiders, and a variety of insects are all known to dig burrows in the arid areas of Argentina, as are many types of mammals. But none of the burrows with which we were familiar resembled those we had found. The burrows we were examining began very near the salt flat, almost in small mounds, but then they seemed to extend up the side of the island to the point where the dense thorn scrub began. Saltbushes grew on the edges of the mounds near the salt flat, and some animal had been foraging on them. We were not seeing the isolated groups of mounds we had expected, the configurations that characterizes both the plains and the golden vizcacha rats. If these burrows were made by an octodontid, it was not a species with which we were familiar.

We marked the spot and returned to camp, where we packed the truck and small trailer and headed back into the salt with Mónica and Rubén. We made camp near the burrows and began to dig. We dug for three days. At night we trapped. Every morning we found that we had caught several interesting animals, including a possible new species of vesper mouse (*Calomys*) and a possible new species of grass mouse (*Akodon*). We also captured a delicate salt mouse (*Salinomys delicatus*), the new genus and species of mouse that Janet and I had named years earlier. It had never been captured so far north, which suggests that it has a very wide distribution, not the rather localized geographic range we had surmised. Our traps also yielded gerbil mice (*Eligmodontia*) and mouse opossums (*Thylamys*). We caught a wide variety of animals, but no vizcacha rat.

After preparing some of the animals as specimens, we would return to dig some more. The burrows seemed almost endless and we moved tons of soil. These burrows seemed to harbor all sorts of life, including tarantulas and another odd spider that we had never seen, almost as large as a tarantula, but a very pale gray in color. We found geckos and other lizards, scorpions, crickets, and even a large, poisonous viper that Janet almost grabbed by the tail, thinking it was the root of a plant. I carefully removed the dark, venomous snake from the burrow and we continued digging. As we dug, we uncovered extensive food stores. In some places saltbush leaves and branches were stored; in others, dense clumps of grass were packed into the burrows, some in large chambers holding perhaps a quart or two of green grass clippings. There was no grass growing in the area as far as we could tell, so these clippings had to have been cut in the summer and tightly packed into the storage chambers, where they remained fresh 3 feet below the surface. We had never seen burrows like these.

After several days we had uncovered no animals and had been unable to capture any in traps. We were exhausted, we were running out of water, and we were thinking about giving up. Maybe the animals had died out on this island. Perhaps they had been picked off by a predator or perished of disease. To make matters worse, the weather was turning nasty. Cold rain was pelting the camp and the temperature was falling. We awoke every morning with a crust of ice over us. If it began to rain steadily we could be isolated in the salt flat for weeks. We did not have enough water for that eventuality, so I decided to break camp and head for the Pipanaco salt flat, where we could do some work on the ecology of the golden vizcacha rat. That day the rain turned to snow and we continued on to Andalgalá, where we holed up for the duration of the bad weather. Four inches of snow fell that night and the temperature plunged below freezing. We had made a wise decision to retreat.

We continued to do fieldwork throughout the Northwest over the next couple of weeks, but I was increasingly bothered by our inability to find the animals that had made those burrows. I decided that we had to try once more to dig them out. We made plans accordingly and returned to the same island three weeks later. We spent three fruitless days digging up burrows of all types, but were unable to find any animals. Our hands bled after being shredded by thorns and our backs ached. We were burned by the sun, which had returned with renewed strength. And eventually, of course, we ran low on water. We were forced to leave. We had brought along a big bag of oranges. When we left, we tossed the remaining dozen oranges on the salt. We were going back to Tucumán, the Garden of the Republic and the major orange-producing region of Argentina. Maybe some of the local fauna could appreciate the oranges we were leaving behind.

Over the next couple of weeks we traveled to high mountain grasslands and high deserts at 15,000 feet of elevation. It was very cold there, and we were preparing animals with frozen hands in the incessantly howling wind and trying to sleep as ice and snow pelted the tents. Through all this discomfort, our failure to find the animal that made the burrows in the Salinas Grandes weighed on my mind. It *had* to be there. We *had* to catch it. We had to return.

I told everyone that we were going back one more time. There were some groans, for it was a long, hard trip. I made it more palatable by saying that we would not dig any more. Clearly, digging was not the way to find this species. The year before I had brought to Argentina some odd traps called Conibear traps. They are designed to trap and kill an animal without the use of bait. You only have to figure out where the animal will be and place the trap in such a way that the animal walks through it. I had used them numerous times, but had never caught a thing in them. I would pin my hopes on these thus far unsuccessful traps. "There must be some reason I carted them down here," I said.

Back we went. The road was now familiar to us. We did not fear the salt. We drove from Tucumán to our old camping spot in seven hours and made camp. I had only a couple of dozen of these unusual traps, so we set them, along with our standard array of traps. There was nothing to do now but wait.

That night I dreamed that I had captured a new octodontid. In my dream, it had entered the Conibear trap only to be dragged away by a predator before I could check the trap in the morning. At three o'clock in the morning I suddenly awoke from the dream. The sun would not be up for hours. I did not think I could find the trap in the dark and it was so cold I did not want to leave the sleeping bag. I decided to go back to sleep and wait for dawn.

I got up at first light and checked the traps. Nothing in the first few traps. Now only two traps remained. Then I saw it. The second to the last trap had an animal in it. It was a vizcacha rat. It appeared to be a new genus and species. We had done it!

On other trips we had saved something special to celebrate the discovery of a new species. Sometimes it was a bottle of champagne, sometimes a cold

Salinas Grandes, Santiago del Estero Province, Argentina. (Photo by M. A. Mares.)

beer on a broiling summer day, or even a can of hearts of palms, considered a delicacy in Argentina. For this trip I had purchased a very expensive bottle of exceptionally good Argentine wine. Our general wine supply had run out long ago, but I had not allowed anyone to open this special bottle, no matter how thirsty or tired we were at the end of the day. It was being saved for a new mammal. "If you want to drink the wine, you have to catch the mammal," I would say.

I hid the new mammal behind my back, hanging it from my belt, and returned to camp. Janet was up already, making coffee. I had assumed a dejected look on my way back.

"Catch anything?" she asked.

"Nothing," I said.

Mounds of the Chalchalero vizcacha rat *(Salinoctomys loschalchalerosorum)*. The animals do not occur in the thorn scrub growing on the body of the island; they live only in the small band of habitat along the periphery. (Photo by M. A. Mares.)

From their respective tents, Rubén and Mónica had also emerged to have coffee.

"Nothing?" they asked.

"Nothing," I said.

The sun was peeking above the horizon and the coffee was boiling. I began to rummage through the food boxes.

"What are you looking for?" Janet and Mónica asked, knowing that I can never find anything around camp.

"I'm looking for that bottle of wine," I said.

Everyone stopped and looked at me. They knew that I had caught an animal and it had to be new.

"Get the wine," I said, and turned around, lifting my jacket so they could see the animal dangling from my belt.

Everyone broke into cheers of joy, with hugs and back slapping all around. We had worked very hard for this animal. We had found it. As it was passed from hand to hand, everyone had comments to make.

"It is a new species!"

"It's a new genus!"

"It looks like it is related to Tympa" (*Tympanoctomys*, the plains vizcacha rat).

"It is really different from Pipa (*Pipanacoctomys*, the golden vizcacha rat).

"I didn't think we'd catch it!"

"I didn't want to come all the way back here."

The new animal looked something like the plains vizcacha rat, but was much larger. It had a narrower head than the golden vizcacha rat, and its tail lacked the big, red brush of the latter. It was mostly brown, and perhaps less striking than the Monte species, but it was new, and it was beautiful.

The Chalchalero vizcacha rat *(Salinoctomys loschalchalerosorum).* (Photo M. A. Mares.)

The sun had returned to the salt flat. We worked there for three more days as the temperature began to climb. The days became extremely hot. We caught one more animal. Then we ran low on water and finally had to eat the oranges that we had discarded several weeks earlier. Most of them tasted fine. We opened cans of fruit and vegetables and drank the surrounding juices. We reached other islands, but found no more burrows. We covered the 5 miles or so of perimeter of the island where we had captured the animals, but again found no burrows. There had to be others, if not on this island, then on more distant islands in the great salt flat. But they would have to wait for another trip. We were out of water and out of time.

I decided to publish the descriptions of the two new genera in the same paper (all four of us were coauthors). Perhaps it was a way of making a statement about the fact that there are still places that harbor unknown organisms, places that need to be explored. I could not remember the last time that two new genera of living mammals were described in the same paper. We named the genus of the second new rodent we discovered after the great salt flat where it occurs (*Salinoctomys*, the octodontid rodent of the salt flat). The species name honors the great Argentine folk group Los Chalchaleros, whose songs we had sung for decades as we camped in a variety of habitats throughout Argentina. They were retiring in 2001, the new millennium, after singing for fifty-two years. We felt that they had accompanied us on every trip we had made over thirty years of field research, so the best way for us to thank them was to name a species of mammal after them. The new octodontid thus became *Salinoctomys loschalchalerosorum*, the Chalchalero vizcacha rat.

We had celebrated our good fortune in finding the new animal by drinking that wonderful bottle of wine, but the joy of discovery was tinged with a hint of melancholy at the thought that this experience would probably never be repeated. Maybe mountain climbers share this feeling when they successfully climb Mount Everest. There is no mountain that is higher that they can ever climb. Others may be more difficult or more dangerous, but all will be compared to the earth's highest peak. No doubt we will continue to talk about the discovery of our new mammals to our colleagues and students, and even embellish the story as the years go by, but each of us knew at the time that we had participated in an extraordinary event. We had predicted new life and found it—and we had done it twice. In a way, we had climbed a pinnacle that we would never climb again.

In our surveys of the area where the golden vizcacha rat lived we also discovered another new species of gerbil mouse, *Eligmodontia*, living on the dunes near the salt flat. That was at the beginning of our late 1999 field season. Before the calendar turned to 2000 our work in the Northwest would lead to the discovery of more than a dozen new species of mammals. In one morning we found three new species. Only six months later we would find the second new genus and species of salt-specialized octodontid, and several other new species of mammals from Andean desert habitats and from the periphery of the great salt flat. We seemed to have gone back in time as we worked our way through the dirt trails of the pre-Andean mountains and valleys of the far Northwest, with new taxa appearing at a rate that was typical of the 1890s, not the 1990s.

Moving from isolated valley to equally isolated mountaintop, from habitat islands at high elevations and other islands at low elevations, each as ecologically distant from the others as the islands in a Pacific archipelago, we continued to encounter totally new or heretofore rare animals. We even found a large population of *Akodon aliquantulus*, the diminutive grass mouse we had described a year earlier and that we thought might be restricted to a grassy mountain "island" in the small province of Tucumán. Crisscrossing the complex terrain of the Northwest, our routes bisected the ancient paths of the Incas, as well as their unexcavated ruins. We gasped for air as we climbed above 15,000 feet, popped Tylenol like candy to ease our headaches, shivered in the freezing Andean winds, and broiled in the heat of the desert lowlands. We were in the field, far from the "real" world—and we could not have been happier.

Land of Diamonds

And of that expedition my final report is that despite its mistakes and its quarrels and its hardships and its dangers, I have enjoyed all of it amazingly . . . It has been a good expedition.

Gordon MacCreagh, *White Waters and Black*, 1926

Each field trip ends with a return to the United States, to a home, to one's "real" job. Invariably, there have been numerous difficulties to overcome in leaving the foreign field site. First, the field trips themselves have been brought to a close. The last trap has been set and the last specimen prepared. It is a time of sadness, and everyone is affected. The sheer loneliness of "shutting down" field research becomes oppressive. Camp is broken for the last time. The tents are aired out and packed.

The mountains, forests, and deserts are left far behind, and the distance is more than merely geographic. One is leaving behind those moments of joy when a new species is discovered, a new insight into the mechanisms of nature is gained, or a cherished hypothesis is modified or discarded altogether in the face of the richness and complexity of life. If the research is part of an ongoing program, a place has to be found to store equipment, specimens, and other materials until they can be unpacked again at some time in the future.

Finally, all that must be done or can be done has been done. Only one more death-defying taxi ride to the airport remains before the plane can be boarded. As it roars down the runway there comes a moment when the heavy plane, its fuel cells laden for the long overnight trip, begins to generate lift as the nose tilts skyward. The wind over the wings lightens the big jet until it leaps into the air when its weight is transferred from the wheels to the wings. It is now a creature of the air.

The country where one has lived in the field, perhaps for a year or more, a country that is now an unforgettable part of one's life, one's history, one's soul, falls away. Usually the plane departs late at night, so it is very dark as it lifts into the sky. Only the lights of the city—Cairo or New Delhi or Buenos Aires—shine brightly against the black landscape. The Andes, or the Sahara, or the rainforest is now only a memory. During the long flight, more than the stars in the sky will change. The world will be different in the morning.

Dawn breaks and red sky fills the windows. Pink turns to blue as ocean and rainforest are crossed. One is swept through the long night on a trip that would have taken Darwin many months at sea. Then the plane enters U.S. airspace. The land looks different. I know the habitats of much of the world from the air, and I always know when I am coming home. The small towns begin to appear. They cannot be confused with the small towns of Europe or Argentina or Australia. Even though the plane is still at 15,000 feet as it descends to land, the towns below welcome me home, proclaiming their difference from the villages and small cities of other lands.

Unlike any other nation, America is a land of diamonds. Baseball diamonds in every town greet the passengers sailing overhead in their ships of the air. Once on final approach to O'Hare Airport in Chicago, I counted fifty baseball diamonds. The pitcher's mound is prominent, as are the foul lines reaching toward infinity—the perfect geometry of the perfect game visible even to God. In this land of baseball, the true extent of the game that helps make our society unique becomes clear. I have returned to the land of baseball, where diamonds reach to the sky and mark the land where the sport was born. The field trip is over and a different society awaits.

I do not sleep in planes, so it is always a long night for me when I travel to a field site or return home. Since I usually work in the Southern Hemisphere, the return trip takes me from summer to winter or winter to summer—and the mirrored seasons of the globe change overnight. I have left South America in suffocating heat and landed in blizzard conditions.

On a Pan Am jet bound from Rio de Janeiro to New York's Kennedy Airport I arrived just as the airport was being closed because of a great blizzard. My plane was forced to fly to Boston, the fuel getting lower by the minute. A second Pan Am jet from Monrovia, Liberia, was also in the air and unable to land at Kennedy. As we both approached Boston, that airport, too, closed because of the great blizzard of seventy-eight that paralyzed the East Coast. Our two planes were in the air after long intercontinental flights and did not have enough fuel to reach an airport that was not socked in with snow.

Logan Airport was opened especially for us and we landed on the snow-covered runways. We passengers in the giant planes could barely see over the massive piles of snow that had been thrown up by the snowplows. We had to be towed to the gates. A day passed before we were finally able to continue our trip, including ten hours spent in an airplane in the snow with little water and no food. As we lifted off the frozen runway on a midnight flight from Kennedy to Pittsburgh, the tropical Brazilian nights faded into a dim memory—even if they had been real a few days earlier. That was fieldwork. Now I was back in the "real world."

The return trip from a field site is often planned to coincide with a holiday, for holidays are the hardest times to be away, especially when one has young children. I have had to travel on Christmas Eve to get home to my sons and share their Christmas—flying all night and arriving in the dawn of Christmas Day like Kris Kringle himself. Once I had promised the boys that I would be home for Halloween, one of their favorite holidays, when neighbors gave them candy and other treats merely because they were wearing costumes that made them seem to be something other than what they were. If only for a night, they became Spider Man or Superman, superheroes in search of a sugar-rich handout. Halloween was a holiday that I had never missed sharing with them. I had always managed to be there to walk them around the neighborhood, enjoying their excitement, protecting them, and even sharing the goodies.

"Will you be here for Halloween, Daddy?" they had asked the summer before when I was leaving for a long field trip. At such moments, I was almost ready to give up fieldwork, for leaving them was almost unbearable. Leaving an adult is hard, but leaving a child is much harder. Children do not grasp the call of your research, or understand why they cannot go along. They only know that you are gone. How much they would grow in the months that I would be away! How much I would miss them each day while I was carrying out research in faraway and lonely places. Would they be okay without me? It was the hardest part of leaving and it never got easier. It is a simple thing to make promises at a time like this, promises that one wants to keep.

"I'll be here," I said. "You can count on it. Would I miss Halloween?" The promise was made. They would remember.

But that field trip was long and hard, and there was a difficult journey back home getting from the field sites to Rio de Janeiro, Brazil, and then getting out of Brazil for the return to the United States. I had used Jeeps, buses, and aged airplanes to arrive at a place where I could transfer to a major airline. Even then, however, it had not been possible to leave on time. The planes had been full. In addition to dangerous weather, there had been mechanical problems and other delays.

I had planned to return the day before Halloween, but events had conspired to make me more than twenty-four hours late. I raced through airports in Miami and New York, desperately trying to get on a plane to Pittsburgh. It was early All Hallows' Eve before I landed at Greater Pittsburgh International Airport, exhausted from the field research and the travel, and suffering from culture shock and jet lag as I adjusted to the frenetic life of the big city.

Life in the field was now a distant memory. I was in limbo—a man without a place. Memories and emotions from opposite hemispheres tugged at me. Portuguese, English, and Spanish crowded together in my mind. I seemed not to fit in the world. This is how I always feel returning home. John Steinbeck in *The Log from the Sea of Cortez* was also overcome with angst on his return from a long voyage of biological exploration: "The real picture of how it had been there and how we had been there was in our minds, bright with sun and wet with sea water and blue or burned, and the whole encrusted over with exploring thought. Here was no service to science, no naming of unknown animals, but rather—we simply liked it. We liked it very much. The brown Indians and the gardens of the sea, and the beer and the work, they were all one thing and we were that one thing too."

Now my time away was over. I had to adjust once again to the rhythm of a different life, where the challenges and the rewards were unrelated to how I had lived for the last several months. With difficulty I put thoughts of the tropics out of my mind. Choruses of frogs and the fluttering wingbeats of bats gave way to the sounds of the city on a holiday.

I hailed a taxi and told the driver that I had to be home before my children went out trick or treating, "and they'll be leaving shortly."

"I don't think you'll make it," he said, "it's getting dark."

"I'll make it. They haven't left yet," I said. "I told them I'd be here and I've come a long way for this. Step on it."

There ensued an unforgettable ride over the 10 miles of busy freeway that lay between the airport and our home. As the cab screeched to a halt in front of the house, Gabriel and Danny, my two Argentines born on different field trips, were just stepping out onto the porch, their goblin outfits on, empty bags in hand. They greeted me as casually as if I had just run down to the corner for some gum. I tossed a $20 bill at the cabby and thanked him.

Without missing a beat, I said to them, "Hey! Let's go trick or treating." I left my bags on the porch and, after having been in transit for three straight days and nights, walked down the street with my boys.

Epilogue

Deserts are among the world's few remaining unmolested wild places. They contain fascinating wildlife, adapted to survival in some of the harshest conditions imaginable. Fears are growing by the day over the potentially disastrous middle- to long-term impact of global warming, and of more land turning to desert not just in the subtropics but further afield. As temperatures rise and rainfall diminishes, never has it been more important for us to understand how deserts function and to conserve the genetic material that exists within them. We simply cannot afford to contribute to the wanton destruction of desert lands and the plants and creatures that dwell in them. The human race ignores desert ecosystems at its peril.

Jim Flegg, *Deserts: A Miracle of Life*, 1993

The earth and its habitats, like living creatures, heal themselves given enough time. Even the massive and worldwide destruction following impacts with celestial bodies—collisions that wiped out most life on the planet—was softened by the gradual forces of geological processes combined with the evolutionary changes of organisms. Tropical rainforest with its teeming and verdant foliage once covered much of the land. With the grand movements of the continental plates in the Middle Cretaceous, 110 million years ago, things began to change. The southern continents formed. Mountains built up as the plates slid over and under one another. Wind patterns changed. World climates diversified.

There were probably always gradients of moisture at different spots on the planet, and some organisms had almost certainly adapted to drier places, but we have no record of them. Evolutionary biology tells us that in such areas, away from the lush forests, organisms would have new opportunities to develop. Centers of new continents and areas located in the shadows of mountains became drier. These areas supported new types of plants: grasses and desert shrubs that were adapted to survive long periods without rain.

Eventually, deserts formed. These arid areas, whether in Africa, Asia, North and South America, or Australia, were the new worlds that remained for organisms to conquer—zones of dryness where few species could thrive. To move into such an extreme environment required a revolutionary evolutionary leap almost as great as that which occurred when life moved from the sea to the land. Plants and animals, bacteria and fungi, vertebrates and invertebrates, all colonized the arid world that had developed over a short period of time, geologically speaking.

It had to have been difficult for organisms associated with abundant moisture throughout most of their history to adapt to life in a region that received water only sporadically, if at all (and my students and I had shown precisely this for mammals in the Brazilian Caatinga). These creatures would have to become very different from their ancestors—more resistant to aridity, less dependant on moisture. Once they entered the zone of aridity they would perforce change greatly. Deserts were a unique place for life to evolve, and many unique forms of life developed in the arid world. Today, many of those singular species are threatened.

Deserts are fragile places—lands where scars last a long time. Fifty thousand years ago a small meteorite struck the desert of Arizona. The crater that resulted from the impact looks remarkably fresh today. There are places in California where you can still see the tracks made by General Patton's tanks when his men prepared for action in North Africa in World War II. Not far away, the tracks of covered wagons heading for the gold fields of California more than a century ago are still visible as parallel dust-blown impressions heading west through the sparse scrublands. The Sahara bears the scars of great tank battles. In Iran, the paths beaten by camels hundreds of years ago still trace the ancient Road of Silk. In the high Andean desert of South America, the signs of the Incas' activities six centuries ago and those of other high desert peoples who lived a millennium before the Incas are still etched across the cold desert plain. There is little soil to support vegetation that would cover these gashes on the desert floor and little rain to wash the tracks away.

Throughout the world, desert habitats are being destroyed, with little thought being given to their unique nature or the incomparable species that inhabit them. Most countries still view the desert as a challenge to be overcome. Perhaps that is because deserts are so harsh and vast. Maybe it is because so few people really understand arid lands. Anyone can drive a modern off-road vehicle into the desert and destroy thousands of organisms. Plants and animals, bacteria and fungi, all maintain a tenuous hold on life in the desert. Each one can be disrupted by a new road being cut, by the clearing of land to build homes, by the diversion of water for agricultural purposes, by deforestation, or by the release of pollutants and other contaminants into the fragile desert system. How can deserts, where even simple tracks can persist for centuries, withstand the constant and deliberate assault of modern society? So far, only their vastness and their harshness have helped deserts resist civilization's onslaught. As regions where survival is all and tenacity the only virtue, deserts remain places of hope. They harbor new things for us to find, new places to explore, and new ways to view the earth and ourselves.

The animals my colleagues and I had discovered on our desert expeditions were biological remnants of old evolutionary events, events that occurred in response to geological, climatological, and botanical upheavals. Dan Janzen and Paul Martin once published a paper dealing with Neotropical anachronisms, species that display characteristics that were coevolved with other species that are now long extinct. They used the example of adaptations of tropical trees that appeared to have evolved as defense mechanisms against being devoured by gomphotheres, extinct relatives of elephants. The gomphotheres are gone, but the trees remain, their life histories reflecting avoidance of predators that are no longer alive. Janzen once referred to such species as the living dead, organisms whose coevolved evolutionary partners—species that were vital for, among other things, pollination or seed dispersal—had gone extinct. Without them, ultimately, the survivors, too, would perish.

In many cases, the populations of the new desert mammals that we are finding are small and isolated—an ideal situation for extinction to occur. A number of these mammals are endangered. One reason no one had ever found them before is that they are rare and isolated. Many of these new animals may well be doomed to extinction within a few years of being described scientifically. They will be known only from the museum specimens, the tissues, and the published reports. Almost no one cares in any way that matters about a rodent living in a desert. To our great sadness, many of our new mammals can also be considered the living dead.

In 1970, when I arrived in Argentina for the first time to work in the Monte Desert, I drove over rough dirt roads that were cross-cut with innumerable channels. Sporadic rains on the surrounding mountains had cut narrow arroyos in the dry soil, and every time the hot air from the lowlands cooled as it rose toward the peaks, it lost precious moisture, causing numerous gullies to form as water rushed down the mountain slopes. Few people lived in the Bolsón de Pipanaco, and those that did were mainly concentrated in the infrequent towns rimming the arid valley. Occasional orchards and a few irrigated fields were limited to a narrow band of green along small rivers that descended from the snow-clad peaks, meandered through the towns, and disappeared into the arid salt wastes of the valley.

The major factors affecting the desert in those days were the cutting down of the mesquite forests for fuel and the large herds of goats that were kept near desert springs. The great hardwood forests that provide blessed food, moisture, and shade for all animals in the desert were in retreat. A desert mesquite lives for hundreds of years, and many of the larger specimens had been old trees when José Francisco de San Martín freed Argentina from Spanish rule in the early 1800s. In a steady-state plant community, one where the forest is neither increasing nor decreasing, each parent tree is replaced by an offspring only after centuries have passed.

Each year the trees flower and fruit, with the pollen, fruit, and leaves providing vital resources for vertebrates and invertebrates alike. But the seeds also needed to be eaten if the trees were to survive, for they had evolved to pass through the gut of a vertebrate before they could germinate. This requirement to have the iron-hard seed coat softened through digestive action was a mechanism assuring dispersal to a new area when the animal dropped the seeds with its excreta, and the starchy and nutritious nature of the seed pods assured they would be eaten by mammals. From the millions of seeds produced by the trees over centuries, only one would reach maturity to replace the parent, and then only after passing through the gut of a mammal.

Over the thirty years I worked in the area I saw many changes, none of them good for the flora and fauna. I was in the Monte when modernity ar-

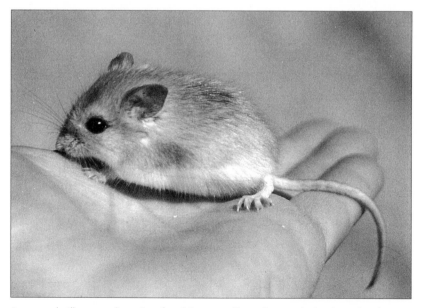

A new and still unnamed species of gerbil mouse, *Eligmodontia*, which is limited to the Bolsón de Pipacaco. (Photo by M. A. Mares.)

A section of the Monte Desert south of Andalgalá, Catamarca Province, Argentina, clear cut and converted to agriculture. The thin white horizontal line across the upper part of the photo is the Pipanaco salt flat, containing the type locality of the golden vizcacha rat. (Photo by M. A. Mares.)

rived in the form of paved roads, new communications systems, and expanded industry and agriculture. The mesquite forests continued to disappear, but much more rapidly. Earlier, people had cut the tough trees into firewood with axes and tied them to their donkeys to transport them to town, but now chain saws fell the largest trees for firewood and parquet flooring, and trucks haul them away. Huge areas of desert are being cleared so that olive orchards and other crops can be planted. Underground water is being used to irrigate the new crops—plants that require much more water than the Monte species they have replaced. Both the desert flora and the desert fauna are in retreat.

In 1972 I collected the first specimen of a new genus of mammal, Olrog's desert mouse (*Andalgalomys olrogi*). This rare little desert mammal lived among the mesquite trees. There are fewer than a dozen of these animals in museum collections and nothing is known of this mouse's ecology. Although we have searched for these mice for the last several years, we have found none. The desert habitat near the type locality—the only place where the species has ever been found—has been greatly changed by destruction of the forest, overgrazing by goats and other livestock, and the heavy use of vehicles in the fragile terrain. I think it is the loss of trees that most affects the species. Has this mammal gone extinct? Certainly nothing was done to protect it, though its unique status has been known for a quarter of a century. Who cares whether a desert mouse shares the world with us?

When we found the new genus of salt-specialized desert rodent in 1999, the golden vizcacha rat (*Pipanacoctomys aureus*), we did so within sight of large machines that were removing the desert vegetation in preparation for planting olive orchards. Billowing smoke from desert plants filled the blue sky as we worked to discover the animal. Drilling rigs dug for underground water even as we dug to find the new mammal. Our uniquely new genus and species, which occurs only over an area of a few square miles, specializes on salt plants, plants whose roots reach into the soil to obtain water from a few feet below the surface. What will happen when the water table begins to drop as underground water is removed and used to irrigate crops? The water level may well sink so deep the salt plant's roots can't reach it. At that point the plants will die, taking with them the new genus and species of rodent that depends on them for its life. The animal is not protected and its habitat is considered to be worthless desert land.

The Chaco, that impenetrable forest, is no longer impenetrable. Soybeans and other crops are replacing enormous areas of woodland. In Tucumán Province most of the thorn forest is gone and rapid habitat conversion continues throughout the region. The Chaco—one of the richest semiarid habitats in South America and the habitat in which the Chacoan peccary, previously thought to have been extinct for ten thousand years, was discovered only twenty-five years ago—has been called the breadbasket of Argentina. The peccary and the brown brocket deer and a host of other species are becoming increasingly uncommon in the area. If you fly over the Chaco, you can see the vast amount of natural habitats that have already been lost— and the pace of change is quickening as national and provincial govern-

ment policies earmark the region for rapid conversion to croplands and grazing lands.

The isolated mountain grassland islands of the pre-Andean chains in
which we found so many new mammals are being subjected to enormous
grazing pressures as road building for mines, colonization, and ski resorts
make it feasible to build ranches and farms in these areas. Potatoes are a favorite crop, and sheep, goats, cattle, and horses find the tough grasses edible.
Fences soon lead to the disappearance of native grasses. The mammals we
have found depend on the grasses and so the small mammals, too, fall back
before the onslaught of development.

Three Andean condors *(Vultur gryphus)* taking wing over the high desert. (Photo by M. A. Mares.)

The puna—that most difficult high Andean desert habitat—has managed
to escape many of the ravages of the lowlands. Its climate is not conducive to
extensive crops, whether or not there is sufficient water. The growing season

Vicuñas in the Laguna Blanca Reserve, Catamarca Province, Argentina. (Photo by M. A. Mares.)

is short and the soils are generally poor. The Incas probably maintained about as many people in the area as could be supported today. Vicuña and guanaco populations have stabilized as hunting has been controlled, but the chinchillas have not returned and many of the new species we discovered are too poorly known for us to speculate about their future. This inhospitable land, though, has used its very uninhabitable nature to remain free of people, abandoned, desertic—*deserare*. This land gives me hope.

Appendix
Selected Readings
Acknowledgments
Index

Appendix

Scientific and Common Names of Species Mentioned in the Text, by Region and Major Category

COMMON NAME	SCIENTIFIC NAME
ARGENTINE BIRDS	
Condor, Andean	*Vultur gryphus*
Duck, Torrent	*Merganetta armata*
Parakeet, Monk	*Myiopsitta monachus*
Parrot, Burrowing	*Cyanoliseus patagonus*
Parrot, Turquoise-fronted Amazon	*Amazona aestiva*
Rhea, Greater	*Rhea americana*
Rhea, Darwin's	*Pterocnemia pennata*
ARGENTINE MAMMALS	
Anteater, Collared	*Tamandua tetradactyla*
Armadillo, Chacoan Fairy	*Chlamyphorus retusus*
Armadillo, Large Hairy	*Chaetophractus villosus*
Armadillo, Pichi	*Zaedyus pichiy*
Armadillo, Pink Fairy	*Chlamyphorus truncatus*
Armadillo, Screaming	*Chaetophractus vellerosus*
Armadillo, Six-banded	*Euphractus sexcinctus*
Capybara	*Hydrochaeris hydrochaeris*
Cavy, Chacoan	*Pediolagus salinicola*
Cavy, Desert	*Microcavia australis*
Cavy, Yellow-toothed	*Galea musteloides*
Chinchilla	*Chinchilla laniger*
Chinchilla Rat, Uspallata	*Abrocoma uspallata*

COMMON NAME	SCIENTIFIC NAME
Deer, Brown Brocket	*Mazama gouazoupira*
Fox, Gray	*Pseudalopex griseus*
Grison, Lesser	*Galictis cuja*
Guanaco	*Lama guanicoe*
Jaguar	*Panthera onca*
Jaguarundi	*Herpailurus yagouaroundi*
Mara	*Dolichotis australis*
Mouse, Bolivian Large Vesper	*Calomys callosus*
Mouse, Burrowing	*Oxymycterus paramensis*
Mouse, Gerbil	*Eligmodontia typus*
Mouse, Grass	*Akodon* (many species)
Mouse, Gray Leaf-eared	*Graomys griseoflavus*
Mouse, Large Vesper	*Calomys venustus*
Mouse, Leaf-eared	*Phyllotis osilae*
Mouse, Diminutive Grass	*Akodon aliquantulus*
Mouse, Morgan's Gerbil	*Eligmodontia morgani*
Mouse, Vesper	*Calomys* (many species)
Mouse, Yellow-nosed Grass	*Akodon xanthorhinus*
Nutria	*Myocastor coypus*
Opossum, Mouse	*Thylamys pallidior*
Opossum, Mouse	*Thylamys venustus*
Opossum, Patagonian	*Lestodelphys halli*
Puma	*Puma concolor*
Rat, Andean	*Andinomys edax*
Rat, Bunny	*Reithrodon auritus*
Sea Lion, South American	*Otaria byronia*
Seal, South American Fur	*Arctocephalus australis*
Tucu-tuco	*Ctenomys* (many species)
Tucu-tuco, Knight's	*Ctenomys knightii*
Vicuña	*Vicugna vicugna*
Vizcacha, Mountain	*Lagidium viscacia*
Vizcacha, Plains	*Lagostomus maximus*
Vizcacha Rat, Brush-tailed	*Octomys mimax*
Vizcacha Rat, Chalchalero	*Salinoctomys loschalchalerosorum*

COMMON NAME	SCIENTIFIC NAME
Vizcacha Rat, Golden	*Pipanacoctomys aureus*
Vizcacha Rat, Plains	*Tympanoctomys barrerae*
Weasel, Patagonian	*Lyncodon patagonicus*

ARGENTINE PLANTS

Apen	*Heterostachys ritteriana*
Cardón	*Trichocereus* (several species)
Creosote Bush	*Larrea cuneifolia*
Mendoza Heliotrope	*Heliotropium mendocina*
Saltbush	*Atriplex lampa*
Vinal	*Prosopis ruscifolia*

ARGENTINE SNAKES

Cascabel (rattlesnake)	*Crotalus durissus*
Yarará (viper)	*Bothrops neuwiedi*

BRAZILIAN MAMMALS

Bat, Flat-headed	*Neoplatymops mattogrossensis*
Cavy, Rock	*Kerodon rupestris*
Fox, Crab-eating	*Cerdocyon thous*
Guinea Pig, Brazilian	*Cavia aperea*
Marmoset, White Tufted-ear	*Callithrix jacchus*
Mouse, Forest	*Bolomys lasiurus*
Opossum, Mouse	*Marmosa karimii*
Opossum, Short-tailed	*Monodelphis domestica*
Punaré	*Thrichomys apereoides*

EGYPTIAN MAMMALS

Bat, Egyptian Fruit	*Rousettus aegyptiacus*
Bat, Tomb	*Taphozous perforatus*
Gerbil, Anderson's	*Gerbillus andersoni*
Gerbil, Fat-tailed	*Pachyuromys duprasi*
Gerbil, Lesser Egyptian	*Gerbillus gerbillus*
Gerbil, Pallid	*Gerbillus perpallidus*

COMMON NAME	SCIENTIFIC NAME
Gerbil, Pleasant	*Gerbillus amoenus*
Jerboa, Four-toed	*Allactaga tetradactyla*
Jerboa, Greater Egyptian	*Jaculus orientalis*
Jerboa, Lesser	*Jaculus jaculus*
Rat, Fat Sand	*Psammomys obesus*

EGYPTIAN SNAKES

Cobra, Egyptian	*Naja haje*
Sidewinder	*Cerastes cerastes*

IRANIAN MAMMALS

Fox, Ruppell's	*Vulpes ruppelli*
Gazelle, Mountain	*Gazella gazella*
Gerbil, Baluchistan	*Gerbillus nanus*
Goat, Persian Wild	*Capra aegagrus*
Hamster, Long-tailed	*Calomyscus bailwardi*
Hedgehog, Long-eared	*Hemiechinus auritus*
Jerboa, Five-toed	*Allactaga elater*
Jerboa, Greater Three-toed	*Jaculus blandfordi*
Jird, Libyan	*Meriones libycus*
Jird, Persian	*Meriones persicus*
Jird, Sundevall's	*Meriones crassus*

NORTH AMERICAN MAMMALS

Antelope Ground Squirrel, White-tailed	*Ammospermophilus leucurus*
Bat, Naked-backed	*Pteronotus davyi*
Bat, Waterhouse's Leaf-nosed	*Macrotus waterhousii*
Bighorn, Desert	*Ovis canadensis*
Bobcat	*Lynx rufus*
Cottontail, Desert	*Sylvilagus audubonii*
Coyote	*Canis latrans*
Deer, Mule	*Odocoileus hemionus*
Fox, Swift or Kit	*Vulpes velox*
Jackrabbit	*Lepus californicus*

COMMON NAME	SCIENTIFIC NAME
Kangaroo Rat, Chisel-toothed	*Dipodomys microps*
Kangaroo Rat, Desert	*Dipodomys deserti*
Kangaroo Rat, Merriam's	*Dipodomys merriami)*
Kangaroo Rat, Panamint	*Dipodomys panamintinus*
Mouse, Brush	*Peromyscus boylii*
Mouse, Cactus	*Peromyscus eremicus*
Mouse, Canyon	*Peromyscus crinitus*
Mouse, Deer	*Peromyscus maniculatus*
Mouse, Northern Grasshopper	*Onychomys leucogaster*
Mouse, Pinyon	*Peromyscus truei*
Mouse, Southern Grasshopper	*Onychomys torridus*
Mouse, Western Harvest	*Reithrodontomys megalotis*
Mouse, White-footed	*Peromyscus leucopus*
Pocket Mouse, Desert	*Chaetodipus penicillatus*
Pocket Mouse, Little	*Perognathus longimembris*
Pocket Mouse, Long-tailed	*Chaetodipus formosus*
Pocket Mouse, Silky	*Perognathus flavus*
Pocket Mouse, Spiny	*Chaetodipus spinatus*
Puma	*Puma concolor*
Rabbit, Pygmy	*Brachylagus idahoensis*
Rat, Cotton	*Sigmodon hispidus*
Squirrel, Ground	*Spermophilus* (many species)
Vole, Sagebrush	*Lemmiscus curtatus*
Woodrat, Desert	*Neotoma lepida*
Woodrat, White-throated	*Neotoma albigula*

NORTH AMERICAN PLANTS

Acacia	*Acacia* (many species)
Big Sagebrush	*Artemisia tridentata*
Cholla	*Opuntia* (many species)
Cottonwood	*Populus* (many species)
Creosote Bush	*Larrea tridentata*
Joshua Tree	*Yucca brevifolia*
Jumping Cholla	*Opuntia fulgida*

COMMON NAME	SCIENTIFIC NAME
Lechuguilla	*Agave lechuguilla*
Mesquite	*Prosopis* (several species)
Ocotillo	*Fouquieria splendens*
Paloverde	*Cercidium* (several species)
Parry's Century Plant	*Agave parryi*
Saguaro	*Carnegiea gigantea*
Saltbush	*Atriplex* (many species)
Soaptree Yucca	*Yucca elata*

Selected Readings

Prologue

Adolph, E. F. 1969. *Physiology of Man in the Desert*. New York: Hafner.

Allan, T., and A. Warren, ed. 1993. *Deserts: The Encroaching Wilderness*. New York: Oxford University Press.

Alloway, D. 2000. *Desert Survival Skills*. Austin: University of Texas Press.

Evenari, M., I. Noy-Meir, and D. W. Goodall, ed. 1985. *Ecosystems of the World*, vol. 12: *Hot Deserts and Arid Shrublands*. New York: Elsevier.

Flores, D. 1999. *Horizontal Yellow: Nature and History in the Near Southwest*. Albuquerque: University of New Mexico Press.

Millet, N. B. 1975. Valley and desert: the two worlds of the Egyptian. In *Man in Nature: Historical Perspectives on Man in His Environment*, ed. L. Levine. Toronto: Royal Ontario Museum.

Schmidt-Nielsen, K. 1972. *How Animals Work*. London: Cambridge University Press.

Young, T. 1972. Population densities and early Mesopotamian origins. In *Man, Settlement, and Urbanism*, ed. P. J. Ucko, R. Tringham, and G. W. Dimbleby. London: Duckworth.

1. The Search for Undiscovered Life

Ghazoul, J., and J. Evans. 2001. Deforestation and land clearing. In *Encyclopedia of Biodiversity*, vol. 2, ed. S. A. Levin. New York: Academic Press.

Heywood, V. H., and R. T. Watson. 1995. *Global Biodiversity Assessment*. Cambridge: Cambridge University Press.

Hoage, R. J., ed. 1985. *Animal Extinctions: What Everyone Should Know*. Washington, D.C.: Smithsonian Institution Press.

King, M. C., and A. C. Wilson. 1975. Evolution at two levels in humans and chimpanzees. *Science*, 188:107–116.

Krumholz, L. R., J. P. McKinley, G. A. Ulrich, and J. M. Suflita. 1997. Confined sub-surface microbial communities in Cretaceous rock. *Nature*, 386:64–66.

Levin, S. A., ed. 2001. *Encyclopedia of Biodiversity*. 5 vols. New York: Academic Press.

May, R. M. 1988. How many species are there on earth? *Science*, 241:1441–1449.

Mittermeier, R. A., N. Myers, P. R. Gil, and C. G. Mittermeier, ed. 1999. *Hotspots: Earth's Biologically Richest and Most Endangered Terrestrial Ecosystems*. Washington, D.C.: Conservation International.

Pace, N. R. 1997. A molecular view of microbial diversity and the biosphere. *Science*, 276:734–740.

Packer, C. 1994. *Into Africa*. Chicago: University of Chicago Press.

Wilson, E. O. 1992. *The Diversity of Life*. Cambridge, Mass.: The Belknap Press of Harvard University Press.

———1994. *Naturalist*. Washington, D.C.: Island Press.

2. *The Immortal Animals*

Beard, P. 1988. *The End of the Game*. San Francisco: Chronicle Books.

Braun, J. K., and M. A. Mares. 1991. Natural history museums: working toward the development of a conservation ethic. In *Latin American Mammalogy: History, Biodiversity, and Conservation*, ed. M. A. Mares and D. J. Schmidly. Norman: University of Oklahoma Press.

Cato, P. S. 1991. The value of natural history collections in Latin American conservation. In *Latin American Mammalogy: History, Biodiversity, and Conservation*, ed. M. A. Mares and D. J. Schmidly. Norman: University of Oklahoma Press.

Cooper, A., C. Lalueza-Fox, S. Anderson, A. Rambaut, J. Austin, and R. Ward. 2001. Complete mitochondrial genome sequences of two extinct moas clarify ratite evolution. *Nature*, 409:704–707.

Edwards, R. Y. 1985. Research: a museum cornerstone. In *Museum Collections: Their Rules and Future in Biological Research*, ed. E. H. Miller. Occasional Papers, British Columbia Provincial Museum, vol. 25.

Genoways, H. H. 1988. Philosophy and ethics of museum collection management. In Zoological Survey of India, *Management of Mammal Collection* [sic] *in Tropical Environment* [sic]. Calcutta: Zoological Survey of India.

Jones, C., W. J. McShea, M. J. Conroy, and T. H. Kunz. 1996. Capturing mammals. In *Measuring and Monitoring Biological Diversity: Standard Methods for Mammals*, ed. D. E. Wilson, F. R. Cole, J. D. Nichols, R. Rudran, and M. S. Foster. Washington, D.C.: Smithsonian Institution Press.

Nicholson, T. D. 1983. The obligation of collecting. *Museum News,* 62:29–33.

Ripley, S. D. 1973. Museums and the natural heritage. *Museum,* 25:10–15.

Rose, C. L., S. L. Williams, and J. Gisbert. 1992. *Current Issues, Initiatives, and Future Directions for the Preservation and Conservation of Natural History Collections.* Madrid: Consejería de Educación y Cultura, Comunidad de Madrid, y Dirección General de Bellas Artes y Archivos, Ministerio de Cultura.

Schlitter, D. A. 1988. The value of recent mammal collections. In Zoological Survey of India, *Management of Mammal Collection* [sic] *in Tropical Environment* [sic]. Calcutta: Zoological Survey of India.

Yates, T. L. 1985. The role of voucher specimens in mammal collections: characterization and funding responsibilities. *Acta Zoologica Fennica,* 170:81–82.

3. Elfin Farmers and Cactophylic Carpenters

Alcock, J. 1985. *Sonoran Desert Spring.* Chicago: University of Chicago Press.

Brown, J. H., and E. J. Heske. 1990. Temporal changes in a Chihuahuan Desert rodent community. *Oikos,* 59:290–302.

Brown, J. H., G. A. Lieberman, and W. F. Dengler. 1971. Woodrats and cholla: dependence of a small mammal population on the density of cacti. *Ecology,* 53:310–313.

Csuti, B. A. 1979. Patterns of adaptation and variation in the Great Basin kangaroo rat (*Dipodomys microps*). *University of California Publications in Zoology,* 111: 1–69.

Darlington, D. 1997. *The Mojave: A Portrait of the Definitive American Desert.* New York: Henry Holt.

Dunbier, R. 1968. *The Sonoran Desert.* Tucson: University of Arizona Press.

Genoways, H., and J. H. Brown, ed. 1993. *Biology of the Heteromyidae.* American Society of Mammalogists, Special Publication no. 10. Lawrence, Kan.: Allen Press.

Grayson, D. K. 1993. *The Desert's Past: A Natural Prehistory of the Great Basin.* Washington, D.C.: Smithsonian Institution Press.

Horgan, P. 1954. *Great River: The Rio Grande in North American History.* New York: Holt, Rinehart and Winston.

Houghton, S. G. 1976. *A Trace of Desert Waters: The Great Basin Story.* Glendale, Calif.: A. H. Clark.

Jaeger, E. C. 1938. *The California Deserts: A Visitor's Handbook.* Stanford: Stanford University Press.

———1957. *The North American Deserts.* Stanford: Stanford University Press.

———1961. *Desert Wildlife.* Stanford: Stanford University Press.

Kenagy, G. J. 1972. Saltbush leaves: excision of hypersaline tissue by a kangaroo rat. *Science*, 178:1094–1096.

Lee, A. K. 1963. The adaptations to arid environments in woodrats of the genus *Neotoma*. *University of California Publications in Zoology*, 64:57–96.

MacMahon, J. A. 1985. *Deserts*. New York: Knopf.

Mares, M. A., ed. 1999. *Encyclopedia of Deserts*. Norman: University of Oklahoma Press.

Murie, J. O., and G. R. Michener. 1984. *The Biology of Ground-dwelling Squirrels*. Lincoln: University of Nebraska Press.

Native Plant Society of Texas. 1996. *The Chihuahuan Desert and Its Many Ecosystems: 1996 Symposium Proceedings, October 17–20, 1995, El Paso, Texas*. Waco, Tex.: Native Plant Society of Texas.

Reichman, O. J., and J. H. Brown, eds. 1983. *Biology of Desert Rodents*. Great Basin Naturalist Memoirs, no. 7. Provo, Utah: Brigham Young University.

Schmidt-Nielsen, K. 1964. *Desert Animals: Physiological Problems of Heat and Water*. New York: Oxford University Press.

———1998. *The Camel's Nose: Memoirs of a Curious Naturalist*. Washington, D.C.: Island Press.

Stones, R. C., and C. L. Hayward. 1968. Natural history of the desert woodrat, *Neotoma lepida*. *American Midland Naturalist*, 80:458–475.

Trimble, S. 1995. *The Sagebrush Ocean: Natural History of the Great Basin*. Reno: University of Nevada Press.

Wauer, R. H., and D. H. Riskind. 1977. *Transactions of the Symposium on the Biological Resources of the Chihuahuan Desert Region, United States and Mexico*. Washington, D.C.: U.S. Department of the Interior, National Park Service, Transactions and Proceedings no. 3.

West, N. E., ed. 1983. *Ecosystems of the World: Temperate Deserts and Semi-deserts*, vol. 5. New York: Elsevier.

4. Darkness and the Cave of the Jaguar

Findley, J. S. 1993. *Bats: A Community Perspective*. Cambridge: Cambridge University Press.

Kunz, T. H., and P. A. Racey, ed. 1998. *Bat Biology and Conservation*. Washington, D.C.: Smithsonian Institution Press.

Packer, C. 1994. *Into Africa*. Chicago: University of Chicago Press.

Wilson, D. E. 1997. *Bats in Question*. Washington, D.C.: Smithsonian Institution Press.

5. The Winding Path to Field Biology

Argentine National Commission on the Disappeared. 1986. *Nunca Más.* New York: Farrar, Straus & Giroux.

Blair, W. F. 1977. *Big Biology.* Stroudsburg, Penn.: Dowden, Hutchinson, and Ross.

Jacobs, M. 1988. *The Tropical Rain Forest: A First Encounter.* New York: Springer-Verlag.

Janzen, D. H., ed. 1983. *Costa Rican Natural History.* Chicago: University of Chicago Press.

MacArthur, R. H. 1972. *Geographical Ecology.* New York: Harper and Row.

Mares, M. A., J. Morello, and G. Goldstein. 1985. The Monte Desert and other subtropical semi-arid biomes of Argentina, with comments on their relation to North American arid areas. In *Ecosystems of the World: Hot Deserts and Arid Shrublands,* ed. M. Evenari, I. Noy-Meir, and D. W. Goodall, vol. 12B. Amsterdam: Elsevier.

Mares, M. A., and D. E. Wilson. 1971. Bat reproduction during the Costa Rican dry season. *BioScience,* 21:471–477.

Morello, J. 1958. La provincia fitogeográfica del Monte. *Opera Lilloana,* 2:1–155.

Rodriguez de la Fuente, F. 1975. *Animals of South America.* London: Orbis.

Seneca. 1971. *Naturales Quaestiones.* Trans. T. H. Corcoran. 2 vols. Cambridge, Mass.: Harvard University Press.

6. The Desert at the Bottom of the World

Branch, L. C. 1993. Social organization and mating system of the plains viscacha *(Lagostomus maximus). Journal of Zoology* (London), 229:473–491.

Cabrera, A., and J. Yepes. 1960. *Mamíferos Sudamericanos.* 2nd ed. 2 vols. Buenos Aires: Ediar.

Darwin, C. 1965 (1839). *The Voyage of the Beagle.* London: Dent.

Flegg, J. 1993. *Deserts: A Miracle of Life.* New York: Blandford Press.

Mabry, T. J., J. H. Hunziker, and D. R. DiFeo, Jr., ed. 1977. *Creosote Bush: Biology and Chemistry of* Larrea *in New World Deserts.* Stroudsburg, Penn.: Dowden, Hutchinson, and Ross.

Mares, M. A., and R. A. Ojeda. 1984. Faunal commercialization and conservation in South America. *BioScience,* 34:580–584.

Ojeda, R. A., and M. A. Mares. 1982. Conservation of South American mammals: Argentina as a paradigm. In *Mammalian Biology in South America,* ed. M. A. Mares and H. H. Genoways. Linesville, Penn.: Pymatuning Laboratory of Ecology, Special Publication no. 6.

Prakash, I., and P. K. Ghosh, ed. 1975. *Rodents in Desert Environments*. The Hague: Junk.

Redford, K. H., and J. F. Eisenberg. 1992. *Mammals of the Neotropics*, vol. 2: *The Southern Cone, Chile, Argentina, Uruguay, Paraguay*. Chicago: University of Chicago Press.

Rodriguez de la Fuente, F. 1975. *Animals of South America*. London: Orbis.

Schmidt-Nielsen, K. 1964. *Desert Animals: Physiological Problems of Heat and Water*. New York: Oxford University Press.

Simpson, G. G. 1980. *Splendid Isolation: The Curious History of South American Mammals*. New Haven: Yale University Press.

Stehli, F., and S. D. Webb, ed. 1985. *The Great American Biotic Interchange*. New York: Plenum.

7. *Different Actors, Different Scripts*

Baker, R. H. 1991. The classification of Neotropical mammals—a historical résumé. In *Latin American Mammalogy: History, Biodiversity, and Conservation*, ed. M. A. Mares and D. J. Schmidly. Norman: University of Oklahoma Press.

Centers for Disease Control and Prevention. 1993. Hantavirus infection—southwestern United States: interim recommendations for risk deduction. *Morbidity and Mortality Weekly Report*, 42 (RR-11):1–13.

Childs, J. E., J. N. Mills, and G. E. Glass. 1995. Rodent-borne hemorrhagic fever viruses: a special risk for mammalogists? *Journal of Mammalogy*, 76:664–680.

Darwin, C. 1965 (1839). *The Voyage of the Beagle*. London: Dent.

Díaz, M. M., R. M. Barquez, J. K. Braun, and M. A. Mares. 1999. A new species of *Akodon* (Muridae: Sigmodontinae) from northwestern Argentina. *Journal of Mammalogy*, 80:786–798.

Gallardo, M. H., J. W. Bickham, R. L. Honeycutt, R. A. Ojeda, and N. Köhler. 1999. Discovery of tetraploidy in a mammal. *Nature*, 401:341.

Garrett, L. 1994. *The Coming Plague: Newly Emerging Diseases in a World out of Balance*. New York: Farrar, Straus & Giroux.

Hershkovitz, P. 1962. Evolution of Neotropical cricetine rodents (Muridae) with special reference to the phyllotine group. *Fieldiana, Zoology*, 46:1–524.

Hill, J. E. 1990. A memoir and bibliography of Michael Rogers Oldfield Thomas, F.R.S. *Bulletin of the British Museum of Natural History (History Series)*, 18:25–113.

Hutchinson, G. E. 1965. *The Ecological Theater and the Evolutionary Play*. New Haven: Yale University Press.

Mabry, T., J. Hunziker, and D. R. DiFeo, Jr., ed. 1977. *Creosote Bush: Biology and Chemistry of Larrea in New World Deserts*. Stroudsburg, Penn.: Dowden, Hutchinson, and Ross.

Mares, M. A. 1975. Observations of Argentine desert rodent ecology, with emphasis on water relations of *Eligmodontia typus*. In *Rodents in Desert Environments*, ed. I. Prakash and P. K. Ghosh. The Hague: Junk.

———1976. Convergent evolution of desert rodents: multivariate analysis and zoogeographic implications. *Paleobiology*, 2:39–64.

———1977. Water balance and other ecological observations on three species of *Phyllotis* in northwestern Argentina. *Journal of Mammalogy*, 58:514–520.

———1977. Water economy and salt balance in a South American desert rodent, *Eligmodontia typus*. *Comparative Biochemistry and Physiology*, 56A:325–332.

———1977. Water independence in a South American non-desert rodent. *Journal of Mammalogy*, 58:653–656.

———1993. Desert rodents, seed consumption, and convergence: the evolutionary shuffling of adaptations. *BioScience*, 43:372–379.

Mares, M. A., J. K. Braun, and R. Channell. 1997. Ecological observations on the octodont rodent, *Tympanoctomys barrerae*, in a saline habitat in south central Argentina. *The Southwestern Naturalist*, 42:488–493.

Mares, M. A., R. A. Ojeda, C. E. Borghi, S. M. Giannoni, G. B. Diaz, and J. K. Braun. 1997. How desert rodents overcome halophytic plant defenses. *BioScience*, 47:699–704.

Mills, J. N., et al. 1998. A survey of Hantavirus antibody in small-mammal populations in selected United States national parks. *American Journal of Tropical Medicine and Hygiene*, 58:525–532.

Orians, G. H., and O. T. Solbrig, ed. 1977. *Convergent Evolution in Warm Deserts*. Stroudsburg, Penn.: Dowden, Hutchinson, and Ross.

Osgood, W. H. 1909. Revision of the mice of the American genus *Peromyscus*. *North American Fauna*, 28:1–285.

Prance, G. T. *Biological Diversification in the Tropics*. New York: Columbia University Press.

Preston, R. 1994. *The Hot Zone*. New York: Random House.

Simpson, G. G. 1965. *The Geography of Evolution*. Philadelphia: Chilton Books.

———1970. The Argyrolagidae, extinct South American marsupials. *Bulletin of the Museum of Comparative Zoology*, 139:1–86.

————1980. *Splendid Isolation: The Curious History of South American Mammals.* New Haven: Yale University Press.

Thomas, O. 1918. A new species of *Eligmodontia* from Catamarca. *Annals and Magazine of Natural History,* 2:482–484.

Williams, D. F., and M. A. Mares. 1978. A new genus and species of phyllotine rodent (Mammalia: Muridae) from northwestern Argentina. *Annals of Carnegie Museum,* 47:193–221.

8. Desert in the Sky

Cabrera, A. L. 1957. La vegetación de la Puna Argentina. *Revista de Investigaciones Agrícolas,* 11:317–342.

————1976. Regiones fitogeográficas Argentinas. *Enciclopedia Argentino de Agricultura y Jardinería,* vol. 1, fasc. 1:1–85.

Carballo, P. M. 1984. *Halcones sobre Malvinas.* Buenos Aires: Ediciones del Cruzamente.

Darwin, C. 1965 (1839). *The Voyage of the Beagle.* London: Dent.

Hastings, H., and S. Jenkins. 1983. *The Battle for the Falklands.* New York: Norton.

Mares, M. A., R. M. Barquez, J. K. Braun, and R. A. Ojeda. 1996. Observations on the mammals of Tucuman Province, Argentina. I. Systematics, distribution, and ecology of the Didelphimorphia, Xenarthra, Chiroptera, Primates, Carnivora, Perissodactyla, and Artiodactyla. *Annals of Carnegie Museum,* 65:89–152.

Mares, M. A., J. Morello, and G. Goldstein. 1985. The Monte Desert and other subtropical semiarid biomes of Argentina, with comments on their relation to North American arid areas. In *Ecosystems of the World,* ed. M. Evenari and I. Noy-Meir. Amsterdam: Elsevier.

Mares, M. A., R. A. Ojeda, J. K. Braun, and R. M. Barquez. 1997. Observations on the mammals of Catamarca Province, Argentina: systematics, distribution, and ecology. In *Life among the Muses: Papers in Honor of James S. Findley,* ed. T. L. Yates, W. L. Gannon, and D. E. Wilson. Albuquerque: University of New Mexico Museum of Southwestern Biology, Special Publication no. 3.

Mayr, E. 1997. *This Is Biology.* Cambridge, Mass.: The Belknap Press of Harvard University Press.

Pond, A. 1962. *The Desert World.* New York: Nelson.

Rodriguez de la Fuente, F. 1975. *Animals of South America.* London: Orbis.

Videla, M. A., and J. A. Suarez. *Mendoza Andina: Precordillera, Altacordillera.* Mendoza, Argentina: Editorial Adaud.

9. The Vampire and the Phantoms of All Hallows' Eve

Mayr, E. 1997. *This Is Biology.* Cambridge, Mass.: The Belknap Press of Harvard University Press.

10. Land of the Shah

Firouz, E. 1974. *Environment Iran.* Tehran: The National Society for the Conservation of Natural Resources and Human Environment.

Lay, D. M. 1967. A study of the mammals of Iran resulting from the Street Expedition of 1962–63. *Fieldiana, Zoology,* 54:1–282.

Mares, M. A., ed. 1999. *Encyclopedia of Deserts.* Norman: University of Oklahoma Press.

Pond, A. 1962. *The Desert World.* New York: Nelson.

Tate, R. 1971. *Desert Animals.* London: Macdonald.

11. Impenetrable Land of Thorns

Barquez, R. M. 1997. Viajes de Emilio Budin: la expedición al Chaco, 1906–1907. *Mastozoología Neotropical, Publicaciones Especiales,* 1:1–82.

Bastien, J. W. 1998. *The Kiss of Death: Chagas' Disease in the Americas.* Salt Lake City: University of Utah Press.

Collier, S., T. E. Skidmore, and H. Blakemore. 1992. *The Cambridge Encyclopedia of Latin America and the Caribbean.* 2nd ed. Cambridge: Cambridge University Press.

Durrell, G. M. 1956. *The Drunken Forest.* New York: Viking.

Mares, M. A., J. Morello, and G. Goldstein. 1985. The Monte Desert and other subtropical semiarid biomes of Argentina, with comments on their relation to North American arid areas. In *Ecosystems of the World,* vol. 12A: *Hot Deserts and Arid Shrublands,* ed. M. Evenari and I. Noy-Meir. Amsterdam: Elsevier.

Mayer, J. J., and P. N. Brandt. 1982. Identity, distribution, and natural history of the peccaries, Tayassuidae. In *Mammalian Biology in South America,* ed. M. A. Mares and H. H. Genoways. Linesville, Penn.: Pymatuning Laboratory of Ecology, University of Pittsburgh.

Miller, E. S., ed. 1999. *Peoples of the Gran Chaco.* Westport, Conn.: Bergin and Garvey.

Rodriguez de la Fuente, F. 1975. *Animals of South America.* London: Orbis.

Wetzel, R. M., R. E. Dubos, R. L. Martin, and P. Myers. 1975. *Catagonus,* an "extinct" peccary, alive in Paraguay. *Science,* 189:379–381.

12. *The Devil's Town*

Collier, S., T. E. Skidmore, and H. Blakemore. 1992. *The Cambridge Encyclopedia of Latin America and the Caribbean.* 2nd ed. Cambridge: Cambridge University Press.

Eisenberg, J. F., and K. H. Redford. 1999. *Mammals of the Neotropics,* vol. 3: *The Central Neotropics: Ecuador, Peru, Bolivia, Brazil.* Chicago: University of Chicago Press.

Lacher T. E., Jr. 1981. The comparative social behavior of *Kerodon rupestris* and *Galea spixii* and the evolution of behavior in the Caviidae. *Bulletin of Carnegie Museum of Natural History,* 17:1–71.

Mares, M. A., and T. E. Lacher, Jr. 1987. Ecological, morphological, and behavioral convergence in rock-dwelling mammals. *Current Mammalogy,* 1:307–348.

Mares, M. A., M. R. Willig, and T. E. Lacher, Jr. 1985. The Brazilian Caatinga in South American zoogeography: tropical mammals in a dry region. *Journal of Biogeography,* 12:57–69.

Streilein, K. E. 1982. Ecology of small mammals in the semiarid Brazilian Caatinga. I. Climate and faunal composition. *Annals of the Carnegie Museum,* 51:79–107.

———1982. Ecology of small mammals in the semiarid Brazilian Caatinga. IV. Habitat selection. *Annals of the Carnegie Museum,* 51:331–343.

Willig, M. R. 1983. Composition, microgeographic variation, and sexual dimorphism in Caatingas and Cerrado bat communities from northeast Brazil. *Bulletin of Carnegie Museum of Natural History,* 23:1–131.

Van Dyke, J. 1901. *The Desert.* Baltimore: The Johns Hopkins University Press.

13. *In the Shadow of the Pyramids*

Arritt, S. 1993. *The Living Earth Book of Deserts.* New York: Reader's Digest Association.

Ayyad, M. A., and S. I. Ghabbour. 1986. Hot deserts of Egypt and Sudan. In *Ecosystems of the World,* vol. 12B: *Hot Deserts and Arid Shrublands,* ed. M. Evenari, I. Noy-Meir, and D. W. Goodall. New York: Elsevier.

Cloudsley-Thompson, J. L., ed. 1984. *Key Environments: Sahara Desert.* New York: Pergamon Press, in collaboration with the International Union for the Conservation of Nature and Natural Resources.

Flegg, J. 1993. *Deserts: A Miracle of Life.* New York: Blandford.

Grenot, C. 1974. Physical and vegetational aspects of the Sahara Desert. In *Desert Biology,* ed. G. W. Brown. New York: Academic Press.

Mares, M. A. 1980. Convergent evolution among desert rodents: a global perspective. *Bulletin of Carnegie Museum of Natural History*, 16:1–51.

Nevo, E., and O. A. Reig, ed. 1990. *Evolution of Subterranean Mammals at the Organismal and Molecular Levels*. New York: Wiley.

Nowak, R. M. 1999. *Walker's Mammals of World*. 6th ed. Baltimore: The Johns Hopkins University Press.

Osborn, D. J., and I. Helmy. 1980. The contemporary land mammals of Egypt (including Sinai). *Fieldiana, Zoology*, 5:1–579.

Seneca, 1971. *Naturales Quaestiones*. Trans. T. H. Corcoran. 2 vols. Cambridge, Mass.: Harvard University Press.

14. Naming the Anonymous

Barquez, R. M., N. Giannini, and M. A. Mares. 1993. *Guide to the Bats of Argentina*. Norman: Oklahoma Museum of Natural History.

Barquez, R. M., M. A. Mares, and R. A. Ojeda. 1991. *The Mammals of Tucuman*. Norman: Oklahoma Museum of Natural History.

Branch, L. C., D. Villarreal, and G. S. Fowler. 1993. Recruitment, dispersal, and group fusion in a declining population of the plains vizcacha (*Lagstomus maximus*; Chinchillidae). *Journal of Mammalogy*, 74:9–20.

Braun, J. K. 1993. Systematic relationships of the tribe Phyllotini (Muridae: Sigmodontinae) of South America. Norman: Oklahoma Museum of Natural History.

Braun, J. K., and M. A. Mares. 1991. Natural history museums: working toward the development of a conservation ethic. In *Latin American Mammalogy: History, Biodiversity, and Conservation*, ed. M. A. Mares and D. J. Schmidly. Norman: University of Oklahoma Press.

————1995. A new genus and species of phyllotine rodent (Rodentia: Muridae: Sigmodontinae) from South America. *Journal of Mammalogy*, 76:504–521.

————1996. Unusual morphologic and behavioral traits in *Abrocoma* (Rodentia: Abrocomidae) from Argentina. *Journal of Mammalogy*, 77:891–897.

Cowan, J. 1996. *A Mapmaker's Dream*. Boston: Shambhala Publications.

Crome, F. H. J. 1997. Researching tropical forest fragmentation: shall we keep on doing what we're doing? In *Tropical Forest Remnants: Ecology, Management, and Conservation of Fragmented Communities*, ed. W. F. Laurance and R. O. Bierregaard, Jr. Chicago: University of Chicago Press.

Hershkovitz, P. 1987. A history of the recent mammalogy of the Neotropical region. In

Studies in Neotropical Mammalogy: Essays in Honor of Philip Hershkovitz, ed. B. D. Patterson and R. M. Timm. Chicago: Field Museum of Natural History, no. 39.

Janzen, D. H. 1996. On the importance of systematic biology in biodiversity development. *ASC Newsletter,* 24:17, 23–28.

Macdonald, D., ed. 1984. *The Encyclopedia of Mammals.* New York: Facts on File.

Mares, M. A. 1988. The need for popular natural history publications concerning mammals. *Proceedings, Workshop on Management of Mammal Collections in Tropical Environments* (Calcutta), 1984:439–452.

———1993. Natural history museums: bridging the past and the future. In *Current Issues, Initiatives, and Future Directions for the Preservation and Conservation of Natural History Collections,* ed. C. L. Rose, S. L. Williams, and J. Gisbert. International Symposium and First World Congress on the Preservation and Conservation of Natural History Collections. Madrid: Ministry of Culture.

Mares, M. A., and J. K. Braun. 1986. An international survey of the popular and technical literature of mammalogy. *Annals of Carnegie Museum,* 55:149–209.

———1996. A new species of *Andalgalomys* (Rodentia: Muridae: Sigmodontinae: Phyllotini) from Argentina. *Journal of Mammalogy,* 77:928–941.

Mares, M. A., R. A. Ojeda, and R. M. Bárquez. 1989. *Guide to the Mammals of Salta Province, Argentina.* Norman: University of Oklahoma Press.

Nowak, R. M. 1991. *Walker's Mammals of the World.* 5th ed. Baltimore: The Johns Hopkins University Press.

Ojeda, R. A., and M. A. Mares. 1989. A zoogeographic analysis of the mammals of Salta Province, Argentina: patterns of community assemblage in the Neotropics. *Special Publications, Texas Tech University* 27:1–66.

Videla, M. A., and J. A. Suarez. *Mendoza Andina: Precordillera, Altacordillera.* Mendoza, Argentina: Editorial Adaud.

Wilson, E. O. 1985. The biological diversity crisis. *BioScience,* 35:700–706.

———1986. The value of systematics. *Science,* 231:1057.

———1992. *The Diversity of Life.* Cambridge, Mass.: The Belknap Press of Harvard University Press.

———1994. *Naturalist.* Washington, D.C.: Island Press.

15. From Howling Wolf Mice to Fairy Armadillos

Arritt, S. 1993. *The Living Earth Book of Deserts.* New York: Reader's Digest Association.

Bowden, C. 1991. *Desierto.* New York: Norton.

Brown, G. W. 1974. *Desert Biology*. New York: Academic Press.

Brown, J. H., O. J. Reichman, and D. W. Davidson. 1979. Granivory in desert ecosystems. *Annual Review of Ecology and Systematics*, 10:201–227.

Darwin, C. 2000. *Charles Darwin's Zoology Notes and Specimen Lists from H.M.S. Beagle*, ed. R. Keynes. Cambridge: Cambridge University Press.

Flegg, J. 1993. *Deserts: A Miracle of Life*. New York: Blandford.

Herrera, J., C. L. Kramer, and O. J. Reichman. 1997. Patterns of fungal communities that inhabit rodent food stores: effect of substrate and infection time. *Mycologia*, 89:846–857.

Mares, M. A. 1980. Convergent evolution among desert rodents: a global perspective. *Bulletin of the Carnegie Museum*, 16:1–51.

———1993. Desert rodents, seed consumption, and convergence: the evolutionary shuffling of adaptations. *BioScience*, 43:372–379.

———1993. Heteromyids and their ecological counterparts: a pandesertic view of rodent ecology. In *The Biology of the Family Heteromyidae*, ed. H. H. Genoways and J. H. Brown. American Society of Mammalogists, Special Publication no. 10. Lawrence, Kan.: Allen Press.

Nowak, R. M. 1991. *Walker's Mammals of the World*. 5th ed. Baltimore: The Johns Hopkins University Press.

Reichman, O. J., A. Fattaey, and K. Fattaey. 1986. Management of sterile and mouldy seeds by a desert rodent. *Animal Behaviour*, 34: 221–225.

Reichman, O. J., and M. Price. 1993. Ecological aspects of heteromyid foraging. In *The Biology of the Family Heteromyidae*, ed. H. H. Genoways and J. H. Brown. American Society of Mammalogists, Special Publication no. 10. Lawrence, Kan.: Allen Press.

Reichman, O. J., and C. Rebar. 1985. Seed preference by desert rodents based on levels of mouldiness. *Animal Behaviour*, 33: 726–729.

16. Aridity's Cornucopia

Arritt, S. 1993. *The Living Earth Book of Deserts*. New York: Reader's Digest Association.

Flegg, J. 1993. *Deserts: A Miracle of Life*. New York: Blandford.

Hoage, R. J., ed. 1985. *Animal Extinctions: What Everyone Should Know*. Washington, D.C.: Smithsonian Institution Press.

Janzen, D. H. 1996. On the importance of systematic biology in biodiversity development. *ASC Newsletter*, 24:17, 23–28.

———.1997. Causes and consequences of biodiversity loss: liquidation of natural capital and biodiversity resource development in Costa Rica. In *Biodiversity and Human Health*, ed. F. Grifo and J. Rosenthal. Washington, D.C.: Island Press.

Mares, M. A. 1992. Neotropical mammals and the myth of Amazonian biodiversity. *Science*, 255:976–979.

Myers, N. 1984. *The Primary Source*. New York: Norton.

17. *Life in the Desert of Salt*

Kuklick, H., and R. E. Kohler, ed. 1996. *Science in the Field. Osiris*, vol. 11.

Mares, M. A., J. K. Braun, R. M. Barquez, and M. M. Díaz. 2000. Two new genera and species of halophytic desert mammals from isolated salt flats in Argentina. *Occasional Papers, Museum of Texas Tech University*, 203:1–27.

Steinbeck, J., and E. F. Ricketts. 1951. *The Log from the Sea of Cortez: The Narrative Portion of the Book "Sea of Cortez."* New York: Viking Press.

18. *Land of Diamonds*

MacCreagh, G. 1926. *White Waters and Black*. 1954. Chicago: University of Chicago Press.

Steinbeck, J., and E. F. Ricketts. 1951. *The Log from the Sea of Cortez: the Narrative Portion of the Book "Sea of Cortez."* New York: Viking Press.

Epilogue

Elliott, D. K., ed. 1986. *Dynamics of Extinction*. New York: Wiley.

Flegg, J. 1993. *Deserts: A Miracle of Life*. New York: Blandford.

Huston, M. A. 1994. *Biological Diversity: The Coexistence of Species on Changing Landscapes*. Cambridge: Cambridge University Press.

Janzen, D. H., and P. S. Martin. 1982. Neotropical anachronisms: the fruits the gomphotheres ate. *Science*, 215:19–27.

Acknowledgments

This manuscript benefited from the editorial suggestions of Michael Fisher at Harvard University Press. I thank him for taking the time to search through the original manuscript to find some value in my story. His gentle comments led me to cut the manuscript by more than half—I only hope the best half remains. I thank Nancy Clemente for excellent editorial suggestions on the entire manuscript. I am also grateful to Patrick Fisher, who answered questions dealing with computerization of figures and text, and John Aguiar, who made editorial suggestions on two chapters.

Several friends, colleagues, and family members were especially helpful during this project. Rubén Barquez, Janet Braun, Tom Lacher, Lynn Mares, Ricardo Ojeda, Laurie Vitt, and Michael Willig read various permutations or portions of the manuscript. Most of my life has been spent with these people and they have lived most of the stories that I could tell.

When one travels to so many countries over so many years it is not possible to thank everyone who has had a part in making research trips successful, but some were so helpful that they require individual recognition. The late Jorge Abalos, Frank Blair, Aristides Leão, Claes Olrog, Abraham Willink, and Vaughan Shoemaker were of great help to me in my earliest days in Argentina and Brazil. Each was an extraordinary person who benefited my life in many ways, large and small. Although all are now gone, none is forgotten. For several decades, Virgilio Roig helped make many of my research projects more successful. Jorge Morello taught me the plants of the Monte Desert at a time when I knew nothing about them. I will always be grateful to both of them. Kamal Wassif in Egypt and Robert Tuck in Iran both made field research in those countries possible. Duane Schlitter took me to Egypt and taught me the fauna of the Sahara Desert. I could not have had a better teacher. I only wish I could have conducted a major study on the marvelous fauna of an Old World desert.

The Martori, Capredoni, and Sanchez families opened their homes to my wife and me in Argentina during the early years, as did Rubén and Patricia Barquez and Dickie and Susana Ojeda for over thirty years. We have shared our lives together. Juan Carlos Schiappa de Acevedo kindly permitted me to study gerbil mice on his estancia in Mendoza Province.

I thank Jim Findley, who saw some spark of promise in a naïve undergraduate, introduced me to mammalogy, and took me on my first field trips—both foreign and domestic. Gene Fleharty taught me to do research and had more faith in me than I had in myself. For help on scientific matters when I was just beginning my research, and for the guiding lights that their own extraordinary careers offered a nascent mammalogist, I thank Charles Lowe, Oliver Pearson, and Michael Rosenzweig.

I am forever grateful to the hundreds of people who have gone with me into the field. Their company and friendship made my life more enjoyable, the work less demanding, and the trips more interesting. We shared a great deal of the best and worst in fieldwork. I especially thank those students and colleagues who worked with me in the field, sometimes for years, often trying to survive the "field trip paranoia" that comes from sharing much time in the field with anyone, even an affable major professor.

My students, now colleagues, Tom Lacher and Michael Willig, were with me on many trips in both North and South America, including intensive research trips in the Brazilian Caatinga and the "wilds" of northwestern Pennsylvania. "Thank you" is hardly sufficient acknowledgment for their friendship and their loyalty. Rubén Barquez and Ricardo Ojeda, first students and now colleagues, worked with me for decades across Argentina, sharing much music, wine, and friendship. The dust we have eaten binds us together always. Janet Braun accompanied me to Argentina on several trips in the 1990s and herself led several expeditions to the most remote habitats. She patiently taught me the complex taxonomy of that country's mammals—including the fact that not all *Akodon* are *Akodon varius*—and made my research papers much better than they would have been without her help. Mónica Díaz worked with me in northwest Argentina for several years, and her singular dedication and youthful energy inspired this aging mammalogist to work harder than I thought I could.

Over the years my research has been supported by the Carnegie Museum, the Ford Foundation, the Fulbright organization (CIES), the National Chicano Council on Higher Education, the National Geographic Society, the National Science Foundation, the Pymatuning Laboratory of Ecology of the University of Pittsburgh, the Sam Noble Oklahoma Museum of Natural History, and the University of Oklahoma. It is a pleasure to acknowledge their help. I also thank the many museums that allowed me to examine the precious specimens kept in their care. They are truly selfless stewards of the world's living heritage. I thank especially the British Museum of Natural History, London; the Carnegie Museum in Pittsburgh; the Colección Lillo and the PIDBA collections of the University of Tucumán and the IADIZA collection in Mendoza, Argentina; the Collection of Mammals at Cairo University; the Collection of Mammals of the University of Arizona in Tucson; the Museo de Ciencias Naturales "Bernardino Rivadavia" in Buenos Aires; the Museum of Natural History in São Paulo, Brazil; the Museum of Southwestern Biology at the University of New Mexico; the Sam Noble Oklahoma Museum of Natural History of the University of Oklahoma; the Transvaal Museum in Pretoria, South Africa; and the United States National Museum of Natural History in Washington, D.C.. Without these museums and their expert curators no one would be able to identify any of the mammals he or she wanted to study.

Finally, my wife, Lynn, made my thesis research possible. She managed the physiology laboratory when I was in the field and accompanied me on many of the early expeditions. She was able to make the field seem like home. More important, in later years I was never concerned about the welfare of my sons, no matter how far away from home I was. I could not have done field research without her help and support. My children, Gabriel Andrés and Daniel Alejandro, also made several field trips with me to the deserts of North America and into parts of the Monte Desert of Argentina. They were of no help with the fieldwork, but they drank a lot of hot chocolate in the early light of dawn and made the field trips much more memorable and enjoyable. They helped me see the wonders of the desert through the eyes of children, and for this, and many other reasons, theirs were the best trips of all.

Old Town
Albuquerque, New Mexico

Index